Optical Interconnects

Optical Interconnects
Ray T. Chen and Chulchae Choi

ISBN-978-3-031-01425-3   paperback
ISBN-978-3-031-02553-2   ebook

DOI 10.1007/978-3-031-02553-2

A Publication in the Springer series
SYNTHESIS LECTURES ON SOLID-STATE MATERIAL AND DEVICES #2

Lecture #2
Series Editor: Sanjay Banerjee, University of Texas at Austin

Series ISSN: 1935-1260   print
Series ISSN: 1935-1724   electronic

First Edition
10 9 8 7 6 5 4 3 2 1

Optical Interconnects

Ray T. Chen
University of Texas at Austin

Chulchae Choi
LG Electronics Institute of Technology
Seoul, Korea

SYNTHESIS LECTURES ON SOLID-STATE MATERIAL AND DEVICES #2

ABSTRACT

This book describes a fully embedded board level optical interconnects in detail including the fabrication of the thin-film VCSEL array, its characterization, thermal management, the fabrication of optical interconnection layer, and the integration of mentioned devices on a flexible waveguide film. All the optical components are buried within electrical PCB layers in a fully embedded board level optical interconnect. Therefore, we can save foot prints on the top real estate of the PCB and relieve packaging difficulty reduced by separating fabrication processes. To realize fully embedded board level optical interconnects, many stumbling blocks need to be addressed such as thin-film transmitter and detector, thermal management, process compatibility, reliability, cost effective fabrication process, and easy integration. The material presented eventually will relieve such concerns and make the integration of optical interconnection highly feasible. The hybrid integration of the optical interconnection layer and electrical layers is ongoing.

KEYWORDS

Fully embedded optical interconnection, vertical cavity surface emitting laser – VCSEL, Optical Interconnection Layer, Printed circuit board (PCB), multichip modules (MCM), very large scale integrated (VLSI) circuits to ultra large scale integrated (ULSI) circuits

Contents

List of Tables

List of Figures

CHAPTER 1

Introduction

1.1   MOTIVATION

The speed and complexity of integrated circuits are increasing rapidly as integrated circuit technology advances from very large scale integrated (VLSI) circuits to ultra large scale integrated (ULSI) circuits. As the number of devices per chip, the number of chips per board, the modulation speed, and the degree of integration continue to increase, electrical interconnects are facing their fundamental bottlenecks, such as speed, packaging, fan-out, and power dissipation. In the quest for high-density packaging of electronic circuits, the construction of multichip modules (MCM), which decrease the surface area by removing package walls between chips, improved the signal integrity by shortening interconnection distances and removing impedance problems and capacitances [1, 2].

The employment of copper and materials with lower dielectric constant materials can release the bottleneck in a chip level for the next several years. The ITRS expects on chip local clock speed will constantly increase to 10 GHz by year 2011. On the other hand, chip-to-board clock speed is expecting slow increasing rate after year 2002 [Fig. 1.1] [3].

The interconnection speed of copper line on printed circuit board cannot run over a few gigahertz. As an example, Fig. 1.2 shows the simulation results of transmission gains for a 7 mils wide and 42 in long copper trace on FR-4 board [4]. The frequency dependent dielectric loss and the skin effect loss dominate the transmission loss. On the other hand, channel noise, resulted from connectors and cross-talk between lines, is increasing with the rate of 20 dB/dec.

For high-fidelity operation, system should secure an adequate signal-to-noise ratio. As a consequence, FR-4 cannot operate beyond a few gigahertz. High-performance materials and advanced layout technology such as IMPS (Interconnected Mesh Power system) are introduced [5]. Especially, IMPS is focused on the signal integrity such as controlled impedance signal transmission with very low cross-talk. The electrical interconnection described by Walker *et al*. provides a 10 Gb/s link over a distance less than 20 m using coaxial cable [6]. However, coaxial cabling is bulky; therefore, it is not suitable for high-density interconnection application.

Current high-performance systems such as parallel computer, ATM machine, switching system, and SAN (Storage Area Network) consist of a number of smaller components and

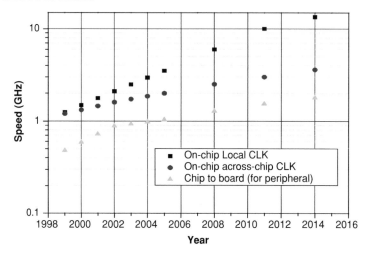

FIGURE 1.1: Interconnection speed roadmap (from ITRS)

require high I/O bandwidth density. The high I/O bandwidth operating at high frequency tends to dissipate more power and be bulky [7]. Electrical interconnects operating at high-frequency region have many problems to be solved such as cross-talk, impedance matching, power dissipation, skew, and packing density. However, there is some hope of solving all of the problems. Power consumption was significantly reduced by adopting low voltage differential signal (LVDS) instead of TTL signal. Typical LVDS has 0.5 V amplitude; hence, power consumption due to the 50 Ω terminal resistor is reduced by 1/50. As the frequency increases, power

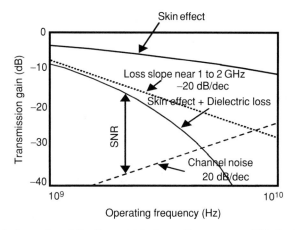

FIGURE 1.2: Transmission gain of 7 mils × 48 in long Cu traces on FR-4 substrate as a function of operating frequency (Courtesy of Accelerant Networks)

consumption due to the terminal resistor becomes trivial because of the increasing resistance of transmission line and the tangential loss of interlayer dielectric materials. Additionally, driver IC should charge and discharge the charge stored in the parasitic capacitor of the transmission line. The high-density interconnect means narrow linewidth and spacing. It increases cross-talk and parasitic capacitance. Therefore, longer electrical interconnection lines with high density consume much more power.

However, optical interconnection has several advantages such as immunity to the electromagnetic interference, independency from impedance mismatch, less power consumption, and high speed operation. Although the optical interconnects have greater advantages compared to the copper interconnection, they still have some difficulties regarding packaging, multilayer technology, signal tapping, and reworkability.

Major performance improvement of a system, so far, comes from the improvements of devices by scaling transistors. The performance improvements of the system will come from new architectures and new technologies, not by relying on incremental reduction of the size of transistors. At present, metal-based electrical interconnects dominate inter- and intra-chip interconnections. The reason for this is simple. There is no viable cost effective alternative. If optical interconnects can provide cost effective solutions, where conventional electrical interconnects fail to function, optical interconnection in a system will be the most promising technology.

1.2 OVERVIEW OF OPTICAL INTERCONNECTS

The history of optical interconnections started from ancient times. At that time, people used to communicate by making a fire and then smoking within the range of our view. At the beginning of electrical communication, telegram and/or telephone drove out light communication due to the convenience of use. Radio wave communication was started after Guglielmo Marconi had succeeded transmitting radio signals and is still popular these days. Increasing data traffic (including voice) and the need for low-cost communication resurrect the optical communication. Long-haul communication was dominated by means of optical fiber.

Several optical interconnect techniques such as free space, guided wave, board level, and fiber array interconnections were introduced for system level applications. Figure 1.3 depicts an example of free space interconnects [8]. A space between two circuit boards or a circuit board and optical interface board is purely empty; so, it is called free space. Light signals coming out from the sources propagate to designated location on the other substrate. The architecture is simple; however, realization is very difficult. All optical components should be mounted at precise location. Moreover, two substrates should be mounted on the designated places exactly. If reflective optical components are used, mounting accuracy should obviously be doubled.

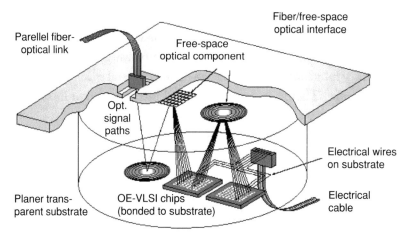

FIGURE 1.3: Illustration of free space optical interconnect (from Gruber [8])

Another disadvantage of the free space interconnects is that the system is vulnerable to external environment such as vibration and dust. Maintenance of the system is also extremely difficult.

Kim and Chen demonstrated bidirectional guided substrate mode optical backplane as illustrated in Fig. 1.4 [9]. Each daughter board has optical transmitters and receivers (TX/RX). The basic architecture of the backplane is a bus structure. All daughter cards share bus lines. The signal coming out of the daughter card is sent to the transparent thick substrate through multiplexed volume hologram. The hologram splits the beam into two directions at an angle of 45°. The beam travels along the substrate by total internal reflection at the interfaces of a substrate. When the beam hits the hologram, some of the light couples out normally and the rest of the light continues propagation. This kind of approach provides bidirectional function unlike any other approaches. However, a number of daughter cards and interconnection density are limited by the divergence of the light. Other issues to be considered are the reliability of

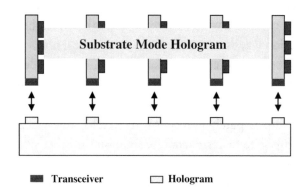

FIGURE 1.4: Illustration of guided substrate mode backplane (from Kim and Chen [9])

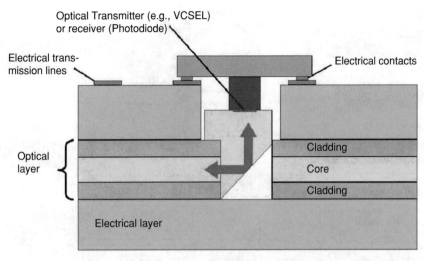

FIGURE 1.5: Illustration of hybrid PCB and coupling from on-board optical device (from Griese [10])

the hologram and alignment. The performance of the hologram tends to be degraded by time elapse, and sensitive to wavelength.

Optical interconnects are still bulky, less reliable, and have packaging difficulties. Many ideas were introduced to overcome aforementioned problems. They are mainly focused on reducing packaging volume and vulnerability to external environment. Scientists use matured semiconductor technology to build optical components. The first effort was fabricating planar lightwave circuits (PLC) on a solid or flexible substrate. Next, PLC was inserted between electrical circuit layers, i.e., hybrid printed circuit board (PCB). Examples of the hybrid integrated PCB are illustrated in Figs. 1.5–1.7 [10–12]. The waveguide layer and electrical layers

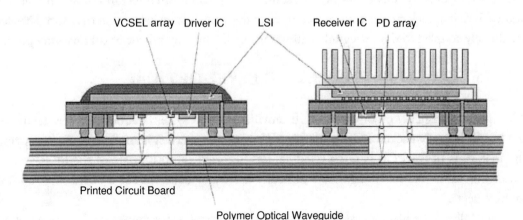

FIGURE 1.6: Illustration of optical-I/O chip packaging concept (from Ishii *et al.* [11])

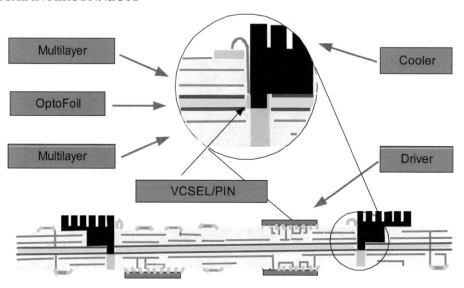

FIGURE 1.7: Concept of electro-optic circuit board (EOCB) (from Krabe *et al.* [12])

are fabricated separately and then laminated together. Optoelectronic devices like VCSELs and PIN-PD are mounted on the top of the hybrid PCB. The optical devices are directly coupled to the waveguide via 45° turning mirror. By doing so, vulnerability can be reduced. However, there is still an alignment problem. After laminating the boards, optoelectronic components should be aligned precisely at designated locations. This procedure is the most difficult one. The end of the waveguide was blocked from the line of sight by the device itself. This alignment diffi-culty can be eased a little by increasing the size of the waveguide. Another shortcoming of these approaches is that the optoelectronic devices occupy real estate of board. The only difference be-tween electro-optic circuit board (EOCB) shown in Fig. 1.7 and architectures shown in Figs. 1.5 and 1.6 is that optoelectronic devices mounted on the heat sink are inserted in the PCB. Devices are directly coupled to the waveguide without using 45° turning mirror or collimation optics.

1.3   FULLY EMBEDDED BOARD LEVEL OPTICAL INTERCONNECTS

The optical waveguides on or within PCB provides a robust system but occupies real estates of the board. Densely packed PCB is used in a high-performance system. Therefore, occupying the footprints of the PCB is not a good choice. Securing the real estate of the PCB, robust packaging, and relief from alignment difficulty are major concerns for a board level optical interconnects.

A fully embedded board-level guided-wave optical interconnection is presented in Fig. 1.8, where all elements, involved in providing high-speed optical communications

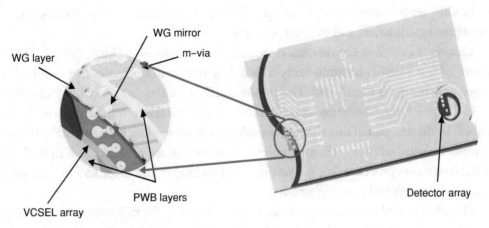

FIGURE 1.8: Illustration of a fully embedded board level optical interconnects

within a board, are shown. These include a vertical cavity surface emitting laser (VCSEL), surface-normal waveguide couplers, and a polymer-based channel waveguide functioning as the physical layer of optical interconnection. The driving electrical signals to modulate the VCSEL and the demodulated signals received at the photoreceiver flow through electrical vias connecting to the surface of the PC board. The fully embedded structure makes the insertion of optoelectronic components into microelectronic systems much more realistic when considering the fact that the major stumbling block for implementing optical interconnection onto high-performance microelectronics is the packaging incompatibility. All the real estate of the PCB surface is occupied by electronics not by optoelectronic components. The performance enhancement due to the employment of the optical interconnection is observed. There is no interface problem between the electronic and optoelectronic components as conventional approaches do.

To realize fully embedded board level optical interconnects, many stumbling blocks need to be addressed such as thin-film transmitter and detector, thermal management, process compatibility, reliability, cost effective fabrication process, and easy integration.

The research work presented herein eventually will relieve such concerns and make the integration of optical interconnection highly feasible.

1.4 CHAPTER ORGANIZATION

Chapter 2 describes the fabrication process of a linear array of the thin-film vertical cavity surface emitting laser (VCSEL). Distributed Brag Reflector (DBR) design theory is addressed. Isolation technique for each device in an array, selective wet oxidation for current confinement, ohmic contacts formation, and finally substrate removal technique are explained.

In Chapter 3, characterization of the VCSEL is described. Characterization includes electro-optical modulation, and thermal property.

Chapter 4 describes the thermal management of a fully embedded VCSEL. The performance of the VCSEL is significantly affected by temperature. Innovative simple thermal management strategy is established. The results of 2D finite element analysis are discussed.

In Chapter 5, the fabrication and calculation of the optical interconnection layer (OIL) are described. In the first part of this chapter, the calculation of coupling efficiency for a 45° waveguide mirror is described. Next, 45° micromirror fabrication process is described. In the last part, the OIL fabrication process using the soft molding technique and deformation compensation technique of the soft mold are explained.

Chapter 6 describes optical layer embodiment including the integration of optoelectronic devices. Integration strategy with printed circuit board is also explained.

Finally, a summary of this research is provided in Chapter 7. Here, our achievement is described.

CHAPTER 2

Thinned Vertical Cavity Surface Emitting Laser Fabrication

2.1 INTRODUCTION

Semiconductor lasers are widely used in optical communications and in optical storage devices. Semiconductor lasers can be categorized into two types on the basis of device structure. One is the edge cleaved laser, and the other is the surface emitting laser. The edge cleaved laser has excellent characteristics. However, the realization of 2D array is nearly impossible. Furthermore, initial probe test is impossible before separation into a chip. In 1977, Iga suggested a vertical cavity surface emitting laser for the purpose of overcoming the aforementioned problems. The fundamental idea was to place a semiconductor gain medium in a Fabry–Perot cavity [13].

Major components of the vertical cavity surface emitting laser (VCSEL) are two mirrors and a gain medium between them. The mirrors are separated by a specific distance for resonance. The mirror is composed of quarter-wave thick stacks which have different refractive indices. The gain medium and mirrors were deposited on the semiconductor wafer by epitaxy.

For a long-haul interconnection, long wavelength (1.3 or 1.5 μm) laser is more suitable due to the low absorption of glass optical fiber. Absorption of the waveguide is not that critical in a short-range interconnection. Therefore, there is no need to use a glass fiber. The electrical-to-optical conversion efficiency and coupling efficiency are more important to reduce power consumption. The VCSEL emitting 850 nm wavelength light is the most suitable source among any other wavelength VCSELs because the manufacturing technology of 850 nm VCSEL is mature and relatively cheap. Furthermore, polymeric waveguide can be used with VCSEL. A very thin VCSEL is needed in a fully embedded board level optical interconnects because the VCSEL is buried between the optical layer and the electrical layers. Thin-film VCSEL design and fabrication process are described in this chapter.

2.2 DESIGN OF AN EPITAXIAL LAYER STRUCTURE

A VCSEL has two distributed Bragg reflectors (DBR), and a cavity. Before designing a DBR, first we have to consider the materials. The materials should have low electrical resistance and low absorption at lasing wavelength. Generally, a 850 nm emitting laser uses $Al_xGa_{1-x}As$ ternary semiconductor.

VSCEL has a small volume of gain medium. To overcome the loss of system, we have to increase feedback by high-reflective mirrors. DBR is composed of many alternative quarter-wave thick stacks. The reflectance of the DBR can be calculated from characteristic matrix of a thin-film stack [14]. If the DBR is composed of q layers, characteristic matrix of the assembly is simply the product of the individual matrices.

$$\begin{bmatrix} B \\ C \end{bmatrix} = \left(\prod_{r=1}^{q} \begin{pmatrix} \cos \delta_r & i \sin \delta_r / \eta_r \\ i\eta_r \sin \delta_r & \cos \delta_r \end{pmatrix} \right) \begin{bmatrix} 1 \\ \eta_m \end{bmatrix} \tag{2.1}$$

$$\delta_r = \frac{2\pi N_r d_r \cos \theta_r}{\lambda}$$

$$\eta_r = y N_r \cos \theta_r \qquad \text{for TE wave}$$
$$\eta_r = y N_r / \cos \theta_r \qquad \text{for TM wave}$$
$$\eta_m = y N_m \cos \theta_m \qquad \text{for TE wave}$$
$$\eta_m = y N_m / \cos \theta_m \qquad \text{for TM wave}$$

where η is the optical admittance of the film or the medium, δ is the phase difference, N_r is the refractive index of the medium, d_r is the physical thickness of the film, λ is the wavelength of light, and θ is the incident angle at the medium. The modified admittance (Y) of film assembly is the ratio of the tangential component of the magnetic field to the tangential component of the electric field.

$$Y = \frac{C}{B} \tag{2.2}$$

The reflectance (R) of the thin-film assembly is calculated by Eq. (2.3).

$$R = \left(\frac{\eta_0 - Y}{\eta_0 + Y} \right) \left(\frac{\eta_0 - Y}{\eta_0 + Y} \right)^* \tag{2.3}$$

The designed VCSEL epi-layer structure is shown in Fig. 2.1. The bottom DBR consisted of 40.5 pairs of $Al_{0.16}Ga_{0.84}As$ and $Al_{0.92}Ga_{0.08}As$. Between $Al_{0.16}Ga_{0.84}As$ and $Al_{0.92}Ga_{0.08}As$ layer, linearly graded index layer was inserted to reduce resistance. The top DBR consisted of 23 pairs of layers and had the same structure as that of the bottom DBR.

The refractive indices of materials used in DBR are summarized in Table 2.1.

The bottom DBR (n-DBR) and the top DBR (p-DBR) were doped with silicon and carbon at the doping level of 10^{18} and 10^{19}, respectively. The calculated reflectances of the

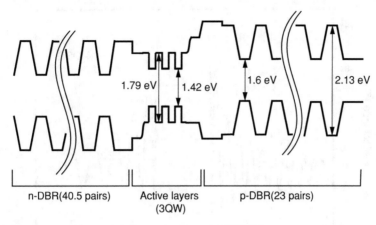

FIGURE 2.1: Energy band diagram of 850 nm emitting VCSEL

DBRs are shown in Fig. 2.2. The reflectances of n-DBR and p-DBR are 99.98% and 99.68% at 850 nm, respectively.

An active layer between DBRs was composed of three GaAs quantum wells. The undoped GaAs epi-layer of 7 nm thickness was surrounded by 10 nm thick $Al_{0.3}Ga_{0.7}As$ layers to form a quantum well. The Fabry–Perot cavity should have the optical thickness of wavelength. To make one wavelength thick cavity, the rest of the space was filled with $Al_{0.6}Ga_{0.3}As$ spacer layer.

$Al_{0.98}Ga_{0.02}As$ layer of 30 nm thickness was inserted between the active and the top DBR layer for selective oxidation to form current aperture. An etch stop layer was also inserted between GaAs substrate and GaAs buffer layer.

The calculated resonance spectrum of VCSEL is shown in Fig. 2.3. In this calculation, material absorption was neglected. The full width at the half maximum (FWHM) of resonance was 0.3 nm.

TABLE 2.1: Refractive Indices of Al_xGa_{1-x} As Materials at 850 nm [15]			
COMPOSITION $AL_XGA_{X-1}AS$	REFRACTIVE INDEX @850 NM	COMPOSITION $AL_XGA_{X-1}AS$	REFRACTIVE INDEX @850 NM
$x = 0$	3.675	$x = 0.6$	3.212
$x = 0.16$	3.505	$x = 0.92$	3.031
$x = 0.3$	3.401	$x = 0.98$	3.000

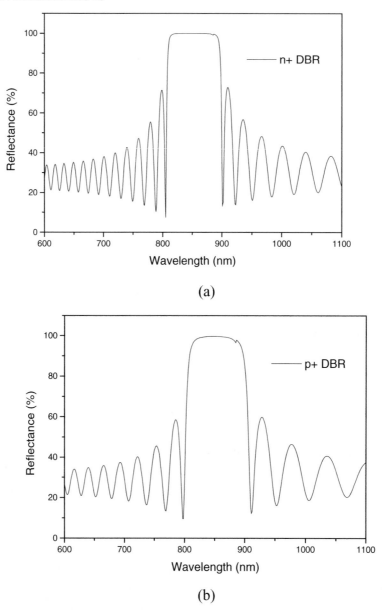

FIGURE 2.2: Calculated reflectance graphs of n-DBR (a) and p-DBR (b)

2.3   FABRICATION OF THE VERTICAL CAVITY SURFACE EMITTING LASER

The epitaxial layers of VCSEL were grown on n-type GaAs wafer using MOCVD at Optowell Co., Ltd. [Table 2.2]. There are two kinds of the current confinement method. One is the proton implantation. The other is the lateral oxide confinement. The proton bombardment forms an

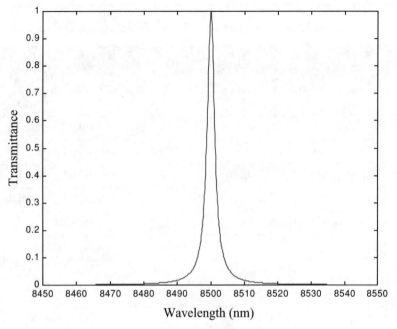

FIGURE 2.3: Calculated resonance curve of designed VCSEL, (neglects material absorption)

insulating region. Therefore, current cannot flow through this region. For the annular-shaped proton bombardment region, a funnel shaped current path is established; therefore, a high-current density region is formed at gain medium. The selective lateral oxidation confinement is based on the difference of oxidation rates according to the material composition. The selective oxidized current confinement method was chosen in my research because of the low-threshold current characteristics and ease of fabrication at our facility.

General thin-film VCSEL process flow is shown in Fig. 2.4. Detailed description will be addressed in the following section.

2.3.1   Device Isolation

The GaAs epi-wafer is electrically conductive; therefore, each device in an array must be isolated. Generally, mesa structure is used for isolation. The annular-shaped trench, which has inner diameter of 42 μm, outer diameter of 80 μm, and depth of 3.2 μm, was formed as shown in Fig. 2.4(a). The depth of the trench was determined to be 3.2 μm by the epi-layer structure because the high-aluminum composition layer for lateral oxidation is located 3.1 μm from the top surface.

First, photoresist AZ5214 was spin coated on the wafer, and annular-shaped patterns were exposed. After developing the patterns, hard bake was carried out at 100°C on a hot plate for 15 min. The epitaxial layers were composed of high and low aluminum composition AlGaAs.

TABLE 2.2: The Epitaxial Layer Structure of a 850 nm VCSEL (from Optowell Co., Ltd.)

LAYER	MATERIAL	COMPOSITION	THICKNESS	DOPING (CM^3)	ACTUAL DOPING	DOPANT
26	GaAs		5	3.00E+19	~4.0E+19	C
25	Al$_x$GaAs	0.16	5	2.00E+19	~2.0E+19	C
24 × 23	Al$_x$GaAs	0.16	41.1	2-3E+18	~2.0E+18	C
23 × 23	Al$_x$GaAs	0.92 -> 0.16	20	2-3E+18		C
22 × 23	Al$_x$GaAs	0.92	49.3	2-3E+18	~3.0E+18	C
21 × 23	Al$_x$GaAs	0.16 -> 0.92	20	2-3E+18		C
20	Al$_x$GaAs	0.16	19.5	2-3E+18		C
19	Al$_x$GaAs	0.98 -> 0.16	20	2-3E+18		C
18	Al$_x$GaAs	0.98	30	2-3E+18		C
17	Al$_x$GaAs	0.92	55.1	2-3E+18		C
16	Al$_x$GaAs	0.6	95.3	N		
15	Al$_x$GaAs	0.3	11	N		
14	GaAs		7	N		
13 × 2	Al$_x$GaAs	0.3	10	N		
12 × 2	Al$_x$GaAs		7	N		
11	Al$_x$GaAs	0.3	11	N		
10	Al$_x$GaAs	0.6	95.3	N		
9	Al$_x$GaAs	0.92	59.6	1-2E+18		Si
8	Al$_x$GaAs	0.16 -> 0.92	20	1-2E+18		Si
7 × 40	Al$_x$GaAs	0.16	41.1	1-2E+18	~1.5E+18	Si
6 × 40	Al$_x$GaAs	0.92 -> 0.16	20	1-2E+18		Si
5 × 40	Al$_x$GaAs	0.92	49.3	1-2E+18	~2.0E+19	Si
4 × 40	Al$_x$GaAs	0.16 -> 0.92	20	1-2E+18		Si
3	Al$_x$GaAs	0.16	10	1-2E+18		Si
2	GaAs		500	1-3E+18	~3.0E+18	Si
1	Al$_x$GaAs	0.98	100	1-3E+18		
0	GaAs	Substrate				

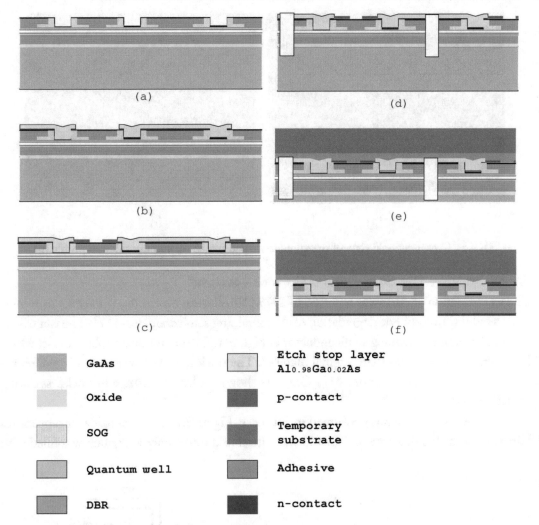

FIGURE 2.4: Thin-film oxide confined VCSEL fabrication flow chart

To etch these layers, nonselective etch solution is required for mesa etching. The sulfuric acid (H_2SO_4) 96%, hydrogen peroxide (H_2O_2) 30%, and water (H_2O) were mixed at the volume ratio of 3:30:150. The etch rate for these epitaxial layers was 1.5 μm/min.

The cross-sectional view of the annular-shaped etched trench is shown in Fig. 2.5. The mesa at the center of the annular-shaped trench will be the VCSEL area. The etchant has an isotropic etch characteristic; therefore, undercut was formed just below the photoresist etch mask. The resultant diameter of the mesa was reduced from 42 μm to about 36 μm. Figure 2.5(b) shows the sidewall profile of etched region. Due to the isotropic etch property of etchant, the sidewall has a curved shape.

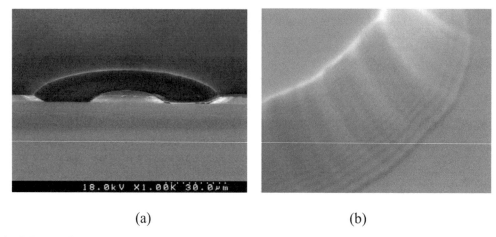

(a)   (b)

FIGURE 2.5: Cross-sectional view of the annular-shaped trench (a) and the sidewall profile (b)

2.3.2   Selective Wet Oxidation for Current Aperture

The wet oxidation of AlAs was reported in 1979 [16]. Researchers at the University of Illinois discovered the atmospheric degradation of Al containing semiconductors [17]. The wet oxidation of aluminum containing semiconductor at high temperature produces robust oxide, which has lower refractive index than original material. The thickness of converted semiconductor is reduced a little during oxidation. $Al_{0.98}Ga_{0.02}As$ shows the least shrinkage in thickness during oxidation.

The wet oxidation was performed in a furnace. Figure 2.6 illustrates oxidation apparatus. Nitrogen gas at the flow rate of 2 L/min was introduced into a bubbler, which was filled with

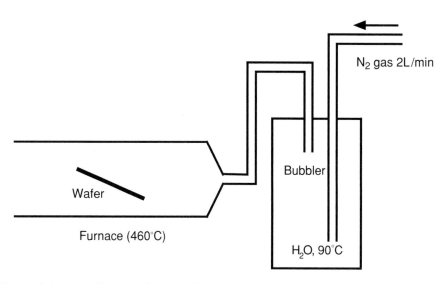

FIGURE 2.6: Schematic diagram of wet oxidation apparatus

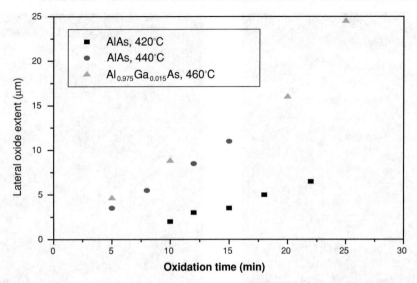

FIGURE 2.7: Lateral oxidation extents as a function of time for AlAs and $Al_{0.975}Ga_{0.025}As$ at different oxidation temperatures

90°C water. Water vapor and nitrogen gas were flown into the furnace at 460°C. The water vapor reacted with AlGaAs and converted AlGaAs to oxide. The mesa etched wafer was placed at tilted angle to form a uniform oxidation over the wafer. The measured lateral oxidation extent of AlAs and $Al_{0.975}Ga_{0.025}As$ at different temperatures are shown in Fig. 2.7. The oxidation rates of AlAs were 0.3 μm/min and 0.73 μm/min at 420°C and 440°C, respectively. The lateral oxidation rate of the $Al_{0.975}Ga_{0.025}As$ was 1.0 μm/min at 460°C. As the aluminum content or oxidation temperature was increasing, oxidation rate was also rising. For all the cases, oxidation rate was constant within a given period.

The selective oxidation to form current aperture is based on different oxidation rates of different aluminum composition $Al_xGa_{1-x}As$ semiconductors. Oxidation layer was composed of 30 nm thick $Al_{0.975}Ga_{0.025}As$. On the other hand, low index layer was composed of $Al_{0.92}Ga_{0.08}As$. The oxidation rate of $Al_{0.975}Ga_{0.025}As$ is four times higher than that of $Al_{0.92}Ga_{0.08}As$ [17]. Therefore, oxide aperture was formed just above the active layer.

The epitaxial layer structure is shown in Fig. 2.8(a). White horizontal line in the picture is the active layer of the VCSEL. Figure 2.8(b) shows the oxidized portion of the annular-shaped trench.

2.3.3   Metallization and Thinning

After the wet oxidation of current aperture layer, annular-shaped trench must be filled with dielectric material to protect further oxidation caused from external environment and to provide isolation from other electrical pads, because the top surface is highly conductive.

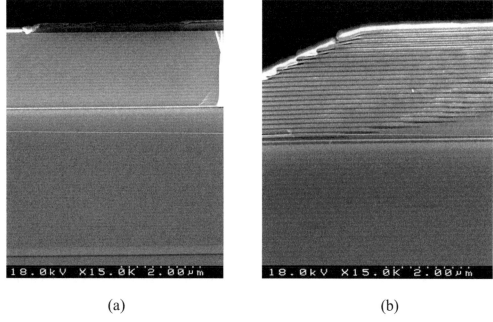

(a)   (b)

FIGURE 2.8: SEM pictures of VCSEL epitaxial layers (a) and cross-sectional view of the oxidized region near the edge of the trench (b)

There are many gap filling dielectric materials such as polyimide, spin on glass (SOG), etc. Photo-imagible polyimide is widely used for isolation purpose. The oxide-confined VCSEL shows less long-term reliability than the ion-implanted VCSEL. It may be caused from the further oxidation of current confinement layer due to prolonged exposure to oxygen and humidity. Polyimide permeates more moisture than SOG. In this research, I chose SOG as a gap filling material. Therefore, I can expect better long-term reliability.

The SOG was spin coated over the wafer at 3000 r/min, and then cured in the furnace at 300°C for 30 min. The thickness of the SOG was 3000 Å after curing. The patterning of the SOG was followed to remove the SOG on the mesa to provide metal contact area. After developing photoresist pattern, the SOG was removed by wet etch using buffered oxide etchant (BOE).

Next process is the metallization of p-contact. Lift-off method was used to define metal contacts. Before depositing metal on the wafer, photoresist (AZ5214) was spin coated, exposed, and developed. The thickness of the photoresist was 1.1 μm on flat surface. However, the thickness of the photoresist on the top of the mesa was only 0.6 μm. It is difficult to lift off metal deposited on a thin photoresist layer. Just before metallization, native oxide formed on the wafer surface was removed by wet etching using high-selective etchant (HCl [1]:H_2O

[1]). Stareev reported extremely low-resistance Ti/Pt/Au ohmic contact to p-type GaAs [18]. Titanium (20 nm thick), platinum (10 nm thick), and gold (100 nm thick) were deposited in sequence in a vacuum chamber. To form a low-resistance ohmic contact, the metals need to be annealed at a high temperature with short-time interval. However, rapid thermal annealing (RTA) process is carried out later just after the deposition of n-contact.

Before thinning wafer, we should consider how to cut extremely thin individual device from the wafer. If a wafer is thick, it provides enough mechanical strength to withstand dicing process. However, if a wafer is thin, it cannot survive the mechanical dicing. Another issue is resulted from reliability after dicing. Mechanical dicing produces lots of damages to the cutting edges. They cause small crack and then finally break the device after integration [19]. To avoid this, a deep chemical etch groove technique, so-called "Dicing-by-Thinning", was introduced [20]. Before thinning the wafer, the outline of the device was deeply etched at the front side of the wafer using chemical etchant. The depth of the etched groove is larger than the thickness of the final device . Therefore, an individual device is separated automatically when thinning was completed.

The fabrication process carried out so far is for a typical thick VCSEL, except for the n-contact formation. However, the thin-film VCSEL process needs additional fabrication steps. The original thickness of GaAs epitaxial wafer is 650 μm. To make very thin VCSEL, we can use either epitaxial lift-off or substrate back etching method.

The epitaxial lift-off (ELO) technique was first developed by Yablovitch [21]; after then, many researchers have succeeded using the technique in various areas such as light emitting diode (LED), VCSEL, MSM photodiode, and double barrier resonant tunneling diode (DBRTD) [22–25]. ELO is based on the high-selective etch of different materials. Hydrofluoric acid has a virtually infinite selectivity to GaAs/AlAs system. The substrates can be removed by dissolving AlAs layer between the substrate and the device layer.

Another approach to make thin-film device is the back etch. The epitaxial layer structure is the same as in ELO technique, but the substrate is removed by etching from the back side. The substrate separated by ELO can be reused, but the back side etch dissolves the substrate. The III-V semiconductor wafer is still expensive compared to a silicon wafer. However, the performance of a device is much important if we consider the possible degradation of performance caused from less excellent surface of a reused wafer. Thus, ELO is interesting, but not widely used these days. Therefore, I used the substrate removal method by back side etching.

Removing 650 μm thick substrate using chemical wet etchant is not realistic. Wet etching can dissolve thick GaAs substrate; however, it produces harmful byproducts. The etch stop layer is about 100 nm to 1 μm, which is very thin. If the thickness of substrate is 650 μm and the etch stop layer is 100 nm thick, then etchant must have a selectivity of 6500 or better. However, there is no etchant with such a high selectivity. Before lapping, the wafer is bonded to a thick glass substrate using the Crystal Bond 509. The bonded wafer is strong enough to survive the harsh lapping environment. Then it is mechanically lapped down to 70 µm by lapping machine, and then chemically etched using medium-selectivity, high-speed etchant, and very high-selectivity etchant, sequentially. The thickness of a mechanically thinned VCSEL wafer was 70 µm including an etch stop layer of 100 nm and 60 µm thick GaAs substrate. It means that the selectivity of an etchant must exceed 600. There is a limitation for using ammonium hydroxide (NH_4OH) and hydrogen peroxide (H_2O_2) system to remove the whole substrate without damaging VCSEL structure. The etch selectivity is not enough; hence, the thin etch stop layer and a part of VCSEL layer will be dissolved unless etching is precisely stopped. The initial etch was carried out using PA etchant. The PA etchant is composed of hydrogen peroxide (H_2O_2) and ammonium hydroxide (NH_4OH) mixed at the volume ratio of 12:1. The PA etchant has selectivity of 35 to $Al_{0.16}Ga_{0.84}As$. The etch rate at ambient temperature was 3 µm/min. Etch was stopped when it reached 30 µm because of uneven thickness during lapping.

Next slow etching was followed by using high-selectivity etchant to remove the rest of GaAs. Generally, citric acid hydrogen peroxide system has high selectivity of ~95 [26]. In our application, this is not enough because of the thin etch stop layer (100 nm) and the relatively thick substrate (60 µm). Therefore, modified citric acid hydrogen peroxide etchant having high selectivity of more than 1000 was used [27].

The etchant is composed of citric acid monohydrate, potassium citrate, and hydrogen peroxide. One mole concentration of citric acid monohydrate in water and the same concentration of potassium citrate were prepared. Just before etching, both solutions were mixed at the

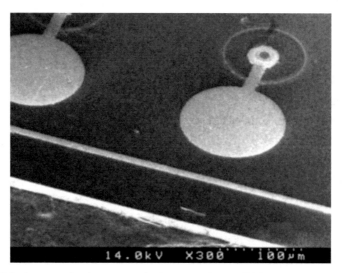

FIGURE 2.9: SEM photograph of a 10 µm thick 1 × 12 linear VCSEL array

same volume ratio. And then, this mixed solution of seven and half part was mixed with fresh one part of hydrogen peroxide. The etch rate of this solution was 240 nm/min for GaAs, and selectivity is over 1000 for $Al_{0.9}Ga_{0.1}As$ and GaAs structure. The etching was stopped when etch stop layer is revealed. The remaining epitaxial layers were VCSEL epitaxial layers (the top and the bottom DBR, and the active layer), the GaAs buffer layer of 50 nm, and the etch stop layer of 100 nm thickness. The etch stop layer must be removed before depositing metal on the GaAs buffer layer. The etch stop layer was completely removed in HF solution. In this process, etching time is not that important because of the nearly infinite selectivity of HF.

The metallization process for n-contact was followed. A gold germanium (AuGe) of 20 nm thickness, nickel (Ni) of 20 nm thickness, and gold (Au) of 120 nm thickness were deposited in sequence. After metal deposition, thin wafer was removed from the glass substrate by raising temperature. The elevated temperature melts a bond material and releases thin VCSEL devices from the glass substrate. The cleaning process was followed to remove any residual on the devices. After cleaning, rapid thermal process (RTP) was carried out at 420°C for 20 s. A fabricated 10 μm thick 1 × 12 linear VCSEL array is shown in Fig. 2.9.

CHAPTER 3

VCSEL Characterization

3.1 ELECTRICAL AND OPTICAL CHARACTERISTICS

Epitaxial layers were grown on 2° off n-GaAs wafer at Optowell Co., Ltd. All data regarding reflectance, photoluminescence, and uniformity were measured at Optowell. The reflectance of the DBR at different areas is shown in Fig. 3.1. Fabry–Perot dip is located at 852 nm. The variation of the Fabry–Perot dip, which indicates uniformity of epitaxial layer, was ±4 nm over 3 in wafer [Fig. 3.2].

The lasing wavelength of the VCSEL at 6 mA was 852 nm and had spectral width (FWHM) of 0.3 nm as shown in Fig. 3.3.

The current–voltage (I–V) characteristic of the VCSEL was measured using Agilent 4155 semiconductor parameter analyzer. Figure 3.4 shows the I–V characteristics of individual VCSEL in the 15 μm aperture VCSEL array. The threshold current was 1.6 V. The deviation of I–V curves resulted in nonuniform oxidized current apertures.

The light-to-current (L–I) characteristics of thick VCSEL were also measured [Fig. 3.5]. The threshold current was 1 mA for VCSEL with 15 μm aperture. Average slope efficiency was 31%. Threshold current density was 566 A/cm^2. The deviation of the L–I characteristic mainly resulted in nonuniform current aperture. Each curve on the graph corresponds to individual VCSEL in an array. Some of them show a thermal roll over at injection current above 5 mA. Uniform oxidation current aperture over the wafer is the most important fact.

The I–V characteristics of the various thick VCSEL is shown in Fig. 3.6. VCSEL was thinned down to 250 μm, and then its I–V characteristic was measured. Next, this VCSEL was thinned down to 200 μm again and measured. Thinning and measuring steps were repeated until whole substrate was completely removed. The diameter of the current aperture was 18 μm. Thinner VCSEL shows low serial resistance than thick device at the same current level. This is caused by reducing thickness of resistive GaAs substrate and better heat dissipation.

The difference of characteristics for VCSELs with various thicknesses is clearly appeared in light-to-current characteristic. L–I characteristics of 250, 200, 250, 100, and 10 μm thick VCSELs are shown in Fig. 3.7. Threshold current density was 393 A/cm^2 corresponding to 3 mA

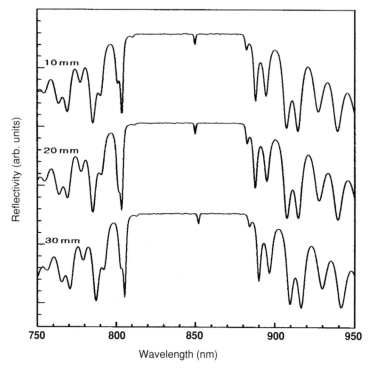

FIGURE 3.1: Measured DBR reflectance graph of various areas on 850 nm VCSEL wafer. [Measured at Optowell Co., Ltd.]

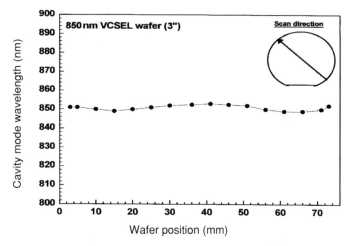

FIGURE 3.2: Epitaxial layer uniformity across the wafer. [Measured at Optowell Co., Ltd.]

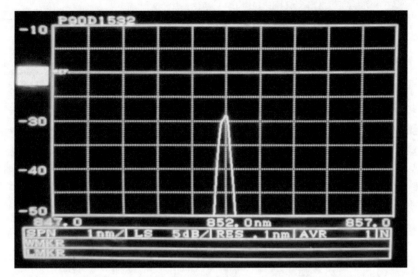

FIGURE 3.3: Emitting spectrum of an 850 nm VCSEL

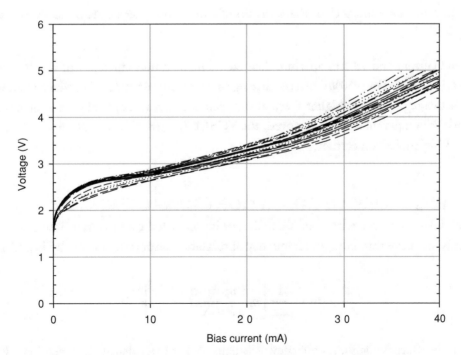

FIGURE 3.4: Current–voltage characteristics of a 15 μm diameter 1 × 12 linear VCSEL array

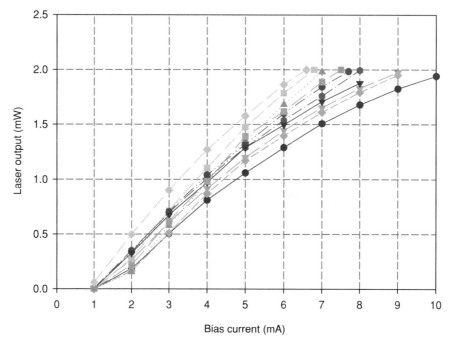

FIGURE 3.5: Light–current (L–I) characteristics of a 15 μm diameter 1 × 12 linear VCSEL array

threshold current and 18 μm aperture diameter. The quantum efficiency of the 10 μm thick VCSEL is increased by ∼50% when the driving current is above 9 mA. The reduced temperature of the active region due to higher thermal conductance increases optical gain in quantum well [28, 29]. When the substrate is removed, the VCSEL (10 μm thick) shows linear dependency even at a high-injection current.

3.2   HIGH-SPEED MODULATION CHARACTERISTICS

The modulation characteristics of VCSEL can be predicted by relaxation resonance. If bias current is far above threshold, the frequency of relaxation resonance is given by Eq. (3.1).

$$f_R = \frac{1}{2\pi}\left[\eta_i \frac{\Gamma v_g}{qV}\frac{\delta g}{\delta N}(I - I_{th})\right]^{\frac{1}{2}} \tag{3.1}$$

where η_i is internal quantum efficiency, g is gain, N is carrier density at given bias, V is the volume of active region, Γ is a confinement factor. From Eq. (3.1), VCSEL can be modulated

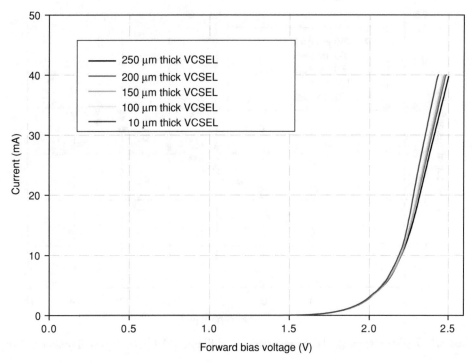

FIGURE 3.6: Current–voltage characteristics of different thickness 18 μm diameter VCSELs

at very high speed because of small active volume. If damping and parasitic capacitance are not large, 3 dB modulation bandwidth is about 1.55 f_R [30]. The next most important factors limiting modulation bandwidth are parasitic capacitance and heat. For high-speed modulation, lower pad capacitance is required.

The measurement of modulation property is a little cumbersome because electrical pads are not coplanar. Bandwidth (4 GHz) probe (Cascade), which has three fingers (GSG) was used to feed drive signal. To test high-speed modulation property, I had to provide low-impedance path. The problem is that the ground contact pad is not on the top surface. Therefore, special biasing is required. Biasing schematic is shown in Fig. 3.8. A VCSEL to be measured is fed by modulated signal. And neighboring VCSELs are tied to the ground. The bottom electrode is biased with negative voltage through an inductor. Two neighboring VCSELs with reverse bias are in a conducting state, i.e., low resistance. Modulated signal drives VCSEL to be measured and flows to lower potential. However, even if bottom electrode is biased with negative potential, modulated current cannot flow through the inductor because the inductor acts like a giant resistor at high frequency. As a result, the current flows through neighboring VCSELs. Thus, the low impedance path is established.

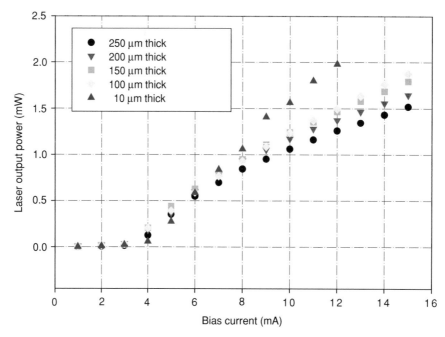

FIGURE 3.7: Light–current (L–I) characteristics of different thickness 18 μm diameter VCSELs

To measure eye diagram, nonreturn-to-zero (NRZ) pulse was directly applied to VCSEL, and DC −2 V was applied to the inductor. Single mode optical fiber was placed close to the VCSEL to tap a light signal. The tapped light was fed to digital oscilloscope through single mode optical fiber. Eye diagram operating at 1 GHz and 2.25 GHz are shown in Fig. 3.9. There was a 100 ps jitter in the eye diagram due to the turn on delay of VCSEL. If high-quality signal and increasing bias voltage are fed to the device, jitter will be decreased.

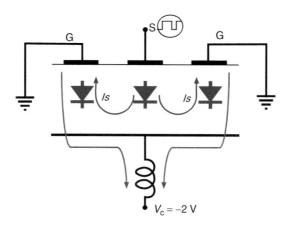

FIGURE 3.8: Biasing schematic for high-speed measurement

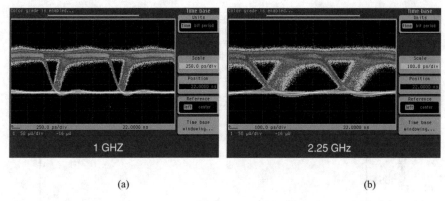

(a)   (b)

FIGURE 3.9: Eye diagrams of modulated VCSEL at 1 GHz (a) and 2.25 GHz (b)

3.3   THERMAL RESISTANCE

The thermal resistance can be calculated from the shift of measured wavelength as a function of substrate temperature and power dissipation. The thermal resistance is given by $R_{th} = \Delta T/\Delta P = (\Delta\lambda/\Delta P)/(\Delta\lambda/\Delta T)$, where ΔT is the change of junction temperature, ΔP is the change of injected power, and $\Delta\lambda$ is the wavelength shift. The junction temperature cannot be measured directly, however, shift of lasing wavelength is related with the junction temperature. Both $\Delta\lambda/\Delta P$ and $\Delta\lambda/\Delta T$ are experimentally confirmed.

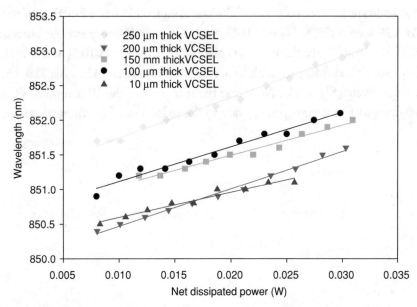

FIGURE 3.10: Wavelength shifts as a function of dissipated power $(\Delta\lambda/\Delta P)$ for different thick VCSEL

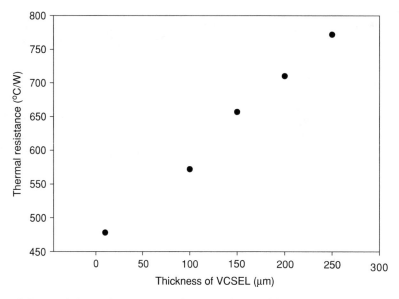

FIGURE 3.11: Measured thermal resistances of various thick VCSELs

The device under test (DUT) was laid on the top of the gallium indium eutectic metal which was used to make an electrical contact. The substrate temperature was controlled by a thermo-electric cooler (TEC).

The measured wavelength shift as a function of temperature ($\Delta\lambda/\Delta T$) for all devices was 0.75 Å/°C at constant power dissipation. The wavelength shifts as a function of net dissipated power ($\Delta\lambda/\Delta P$) were 0.59, 0.54, 0.5, 0.43, and 0.36 Å/mW, respectively, corresponding to 250, 200, 150, 100, and 10 μm thick VCSELs as indicated in Fig. 3.10. The thermal resistances for 250, 200, 150, 100, and 10 μm thick VCSELs were measured to be 772, 710, 657, 572, and 478°C/W, respectively [Fig. 3.11]. Note that the thickness of the 10 μm thick VCSEL has an exclusive advantage of heat management due to the reduction of the thermal resistance.

CHAPTER 4

Thermal Management of Embedded VCSEL

4.1 INTRODUCTION

All the optical components such as light sources, channel waveguides, waveguide couplers, and detectors are inserted within electrical layers in a fully embedded board level optical interconnection as indicated in Fig. 1.8. However, in this configuration, the VCSEL array raises thermal management concerns because it is encapsulated with thermal insulators such as polymer waveguides and bonding film (prepreg). Only the common bottom metal contact of the VCSEL array can be used as a thermal interface. The improper heat dissipation can lead to thermal runaway. The increasing temperature leads to wavelength shift, increasing threshold current, reducing quantum efficiency, shortening device life time, and dissipating more power. Therefore, the heat management of the driving VCSEL array is a critical issue in the fully embedded structure. Lee *et al.* reported the thermal management of a VCSEL-based optical module [31] and Rui Pu *et al.* reported the thermal resistance of a VCSEL bonded to an IC [32]. Both papers presented valuable results, but not applicable to our architecture.

Following section describes the simulated results of the thermal resistance of the VCSEL fully embedded in the PCB as a function of the VCSEL's thickness, and also determines the effective heat sink structure.

4.2 TWO DIMENSIONAL FINITE ELEMENT ANALYSIS

The VCSEL is a major heat source in a fully embedded guided-wave optical interconnects. The embedded VCSEL arrays are thermally isolated by surrounding insulators; therefore, heat builds up and the operating temperature increases. The high-operating temperature may reduce life time of the device and the laser output power. The reliable operation of the VCSEL is accomplished through proper heat management. An effective heat removal is a challenging task in the embedded structure because I have to consider packaging compatibility to the PCB manufacturing process while providing an effective and simple cooling mechanism.

The innovated heat management system for the fully embedded approach is introduced. The key idea is using an n-contact metal affiliated with the bottom DBR mirror of the VCSEL die as a heat spreader, and a part of the heat sink by directly electroplating with copper during the integration process. In general, thermally conductive paste (usually contains fine metal flakes) was used for low-power laser diode packaging. In a high-power laser diode packaging, the gold–tin (Au–Sn) eutectic alloy was used to bond LD on heat sink. The thermally conductive paste has ten times smaller thermal conductivity than that of copper. Therefore, thermally conductive paste cannot be used because it has lower thermal conductivity than copper.

Usually, several tens of micrometer thick copper was deposited in copper contained acid chemical solution during the PCB process. It can be used as a very good electrical and thermal passage, simultaneously.

The thermal resistance of a buried VCSEL depends on the device structure and the packaging structure. Once the device was designed, there is no other way to change thermal resistance of the device. The direct bonding of the device using Cu electroplating reduces thermal resistance of the device due to the absence of the lower thermal conductivity bonding epoxy.

Thermal resistance can be calculated by solving thermal diffusion equation [Eq. (4.1)].

$$\frac{\partial T}{\partial t} = \frac{k}{\rho C_{p}} \nabla^2 T + \dot{q} \frac{1}{\rho C_{p}} \qquad (4.1)$$

Here, T is the temperature, k is the thermal conductivity, ρ is the material density, C_{p} is the specific heat, and $\dot{q}$ is the heat generation rate. For a steady state and a constant heat generation case, Eq. (4.1) turns into a simple Poisson's equation [Eq. (4.2)].

$$k \nabla^2 T + \dot{q} = 0 \qquad (4.2)$$

Once the temperature difference (ΔT) between the device and heat sink was known, thermal resistance (R_{th}) of the embedded VCSEL can be calculated by Eq. (4.3).

$$R_{th} = \Delta T / \Delta P \qquad (4.3)$$

Here, ΔP is the dissipated power.

ANSYS software was used to perform a 2D finite element thermal distribution analysis. The thermal conductivities of GaAs, DBR mirror, and copper are 4.6×10^{-5}, 2.3×10^{-5}, and 4×10^{-4} W/μm·K, respectively [33, 34] [Table 4.1]. As VCSEL parameters, the active diameter of 18 μm and the thickness of 0.3 μm were used.

Heat is generated due to the Bragg reflector's resistance and imperfect conversion efficiency in the active region. However, the heat generated due to DBR is relatively small compared with the active region, therefore, I ignored this term in our simulation [32]. The heat generation rate in the active region (circular shape, diameter of 18 μm) is based on the measured value which is 20 mW per VCSEL.

We compared two different cooling structures as depicted in Fig. 4.1. One is the 250 μm thick bulk copper as a conductive material [Fig. 4.1(a)] and the other is a 30 μm thick

TABLE 4.1: Thermal Conductivities of Materials.

MATERIAL	THERMAL CONDUCTIVITY [W/μM K]	MATERIAL	THERMAL CONDUCTIVITY [W/μM K]
GaAs	4.6×10^{-5}	Au	3.2×10^{-4}
AlGaAs DBR	2.3×10^{-5}	Au/Sn(80:20)	6.8×10^{-5}
H20E (Thermal paste)	2.9×10^{-5}	Si	1.5×10^{-4}
Copper	4×10^{-4}	FR-4	1.7×10^{-7}

electrodeposited copper film [Fig. 4.1(b)]. Producing the 250 μm thick copper in real packaging is nearly impossible. To overcome the realization of thick heat sink, the electroplated copper foil on the back side of VCSEL was introduced. I chose the 30 μm thick copper film as a heat sink because this is the thickness of the copper trace in the electrical layer of the PCB. The copper film can be directly electro-deposited on the n-contact metal pad (Au-Ge/Ni/Au) of the VCSEL array during the electroplating step. The copper foil heat sink looks like a rectangular cap. During simulation the bottom surface of the copper block or the thermal conductive paste or the copper foil is maintained at 25°C.

Figure 4.2 is the automatically generated triangular mesh profile for 10 μm thick VCSEL on the copper foil. The simulation results are shown in Figs. 4.3–4.6. Note that the bottom portion of the drawing was truncated in Fig. 4.2.

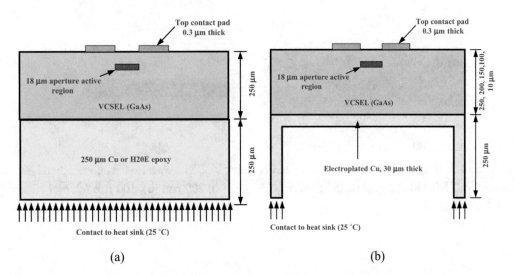

(a)   (b)

FIGURE 4.1: Two types of embedded heat sink structure, electroplated copper or thermal conductive paste (250 μm thick) (H20E, Epotek) (a) and electroplated copper film (30 μm thick) (b)

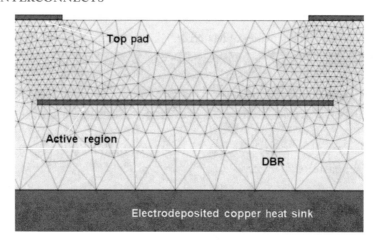

FIGURE 4.2: Generated 2D mesh structure for 10 μm thick VCSEL and 30 μm thick electroplated copper film

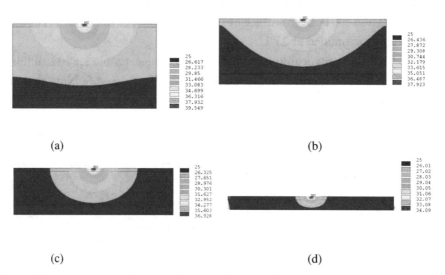

FIGURE 4.3: Temperature distributions of various thick VCSELs with In–Ga eutectic metal interface between VCSEL and the gold plated Si-wafer, 250 μm (a), 200 μm (b), 100 μm (c), and 10 μm (d) thick VCSELs

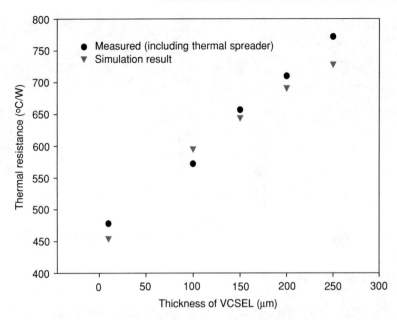

FIGURE 4.4: Comparison of the measured and the calculated thermal resistances for various thick VCSELs. [Active area: diameter of 18 μm, height of 0.3 μm]

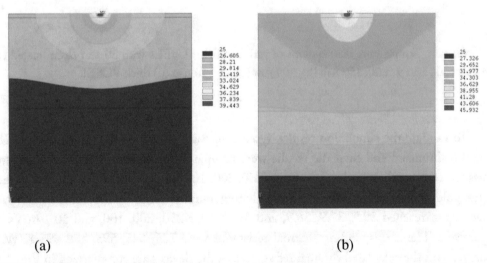

FIGURE 4.5: Temperature distribution of 250 μm thick VCSEL. 250 μm thick electroplated copper heat sink (a) and 250 μm thick thermal conductive paste heat sink (b). [H20E, Epotek]

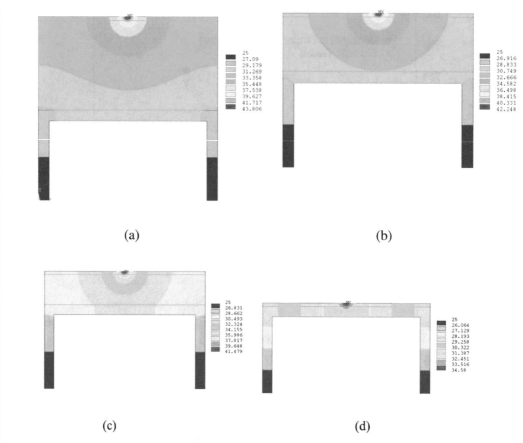

(a)   (b)

(c)   (d)

FIGURE 4.6: Temperature distributions of various thick VCSELs with 30 μm electroplated copper film heat sink, 250 μm (a), 150 μm (b), 100 μm (c), and 10 μm (d) thick VCSEL

To validate the simulation results, typical measuring setup with various thick VCSELs were also simulated and then the results were compared to the measured results. Figure 4.3 shows temperature distribution of various (250, 200, 100, and 10 μm) thick VCSELs attached on the gold coated silicon wafer using indium–gallium eutectic metal. Temperatures at the active region reached 39.5, 37.9, 36.9, and 34.9°C for 250, 200, 100, and 10 μm VCSEL, respectively. The corresponding thermal resistance were 725, 645, 595, and 495 K/W. The measured and the calculated thermal resistances of the devices are summarized in Fig. 4.4. As shown in Fig. 4.3, the calculated thermal resistances of the devices are well matched with the measured results. According to this result, the simulation model and the process were carried out properly.

For the 250 μm thick copper heat sink block, the temperature at the active region reached 39.4°C corresponding to a thermal resistance of 722 K/W [Fig. 4.5(a)]. Despite of lower thermal resistance of the 250 μm thick copper heat sink block, this structure is not realistic in a fully embedded structure due to the difficulty in producing this thickness by electroplating.

For 250 μm thick thermal conductive paste instead of copper block, junction temperature reaches to 45.9°C corresponding to a thermal resistance of 1045 K/W which is 44% higher than that of a copper block. Due to the high-thermal resistance of an integrated VCSEL, thermal conductive paste cannot be used for a fully embedded application.

For the case of a 30 μm thick electrodeposited copper film heat sink with 10 μm thick VCSEL, the junction temperature reached 34.58°C as shown in Fig. 4.6(d) corresponding to the thermal resistance of 455 K/W. The higher junction temperature reduces the quantum efficiency and finally causes catastrophic failure of the device. Figure 4.7 shows theoretically determined thermal resistances for VCSELs with various thicknesses. For a 30 μm thick electroplated copper film, the junction temperatures were theoretically determined to be 43.8, 43, 42.2, 41.5, 40.2, and 34.6°C for 250, 200, 150, 100, 50 and 10 μm thick VCSEL, respectively. The substrate-removed VCSEL having a total thickness of 10 μm shows superior optical and thermal characteristics.

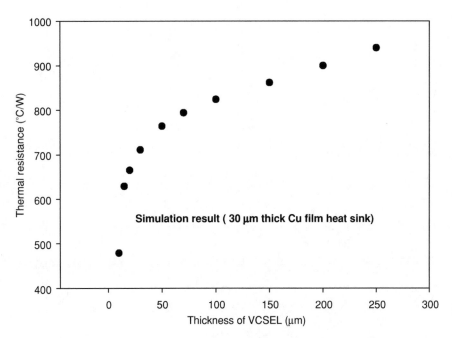

FIGURE 4.7: Calculated thermal resistances of various thick embedded VCSELs

Although the substrate-removed VCSEL has superior properties, fabrication and handling are very difficult. If I assume that thermal resistance of the normal thickness VCSEL (250 μm thick) packaged on bulk copper heat sink is a reference for reliable operation, maximum thickness of an embedded VCSEL will be 70 μm according to the simulation results. Tens of micrometer thick devices can be easily fabricated. The handling of the device is also much easier.

CHAPTER 5

Optical Interconnection Layer

5.1 INTRODUCTION

The optical interconnection layer (OIL) is a key component in fully embedded board level optical interconnects. Due to the fully embedded structure in the PCB, optical waveguide film must be thin and flexible. Furthermore, light source (VCSEL), and photodetector also should be thin enough. Many problems still remain to be solved.

1. Compatibility with industrial standard PCB process;
2. Low cost material and process;
3. Manufacturability; and
4. Reliability.

The compatibility with the PCB process means that the OIL should be treated like a traditional electrical layer and survives harsh PCB lamination environments. Molding process or imprinting is the most suitable method for low-cost approach and manufacturability. The materials of the OIL should have well-matched thermal expansion coefficients to prevent delamination. The integration of the VCSEL and the photodetector onto the waveguide film should be easy and robust. And the integrated OIL must provide good heat sink or spreader for reliable operation of the VCSEL.

In this chapter, micromirror coupler and the fabrication technique of flexible waveguide will be explained.

5.2 WAVEGUIDE MICROMIRROR COUPLER

To efficiently couple optical signals from vertical cavity surface emitting lasers (VCSELs) to polymer waveguides and then from waveguides to photodetectors, two types of waveguide couplers are investigated. They are tilted grating couplers and 45° waveguide mirrors. There are a large number of publications in grating design [35–39]. However, the surface–normal coupling scenario in optical waveguides has not been carefully investigated so far. The profile of tilted grating greatly enhances the coupling efficiency in the desired direction. The phenomenon of grating-coupled radiation is widely used in the guided-wave optical interconnects. Very often, coupling in a specific direction is required. To achieve this unidirectional coupling, the tilted grating profile is selected as a high-efficiency coupler. A very important aspect of manufacturing of such coupler is the tolerance interval of the profile parameters, such as the tooth height, the width, and the tilt-angle. However, the tilted grating coupler has inherent wavelength sensitivity and is not applicable for planarized waveguide.

The 45° waveguide mirror coupler is a very critical component in optical interconnection applications especially in the planarized lightwave circuits (PLC). The mirror can be incorporated with a vertical optical via to enable 3D optical interconnects and couples light to the waveguide. The 45° waveguide mirror is insensitive to the wavelength of light and has high coupling efficiency. There are various techniques to fabricate 45° mirror such as laser ablation [40], oblique reactive ion etching (RIE) [41], temperature controlled RIE [42], reflow [43], and machining [44]. The laser ablation method is subjected to lower throughput and surface damage. The oblique RIE method is limited by directional freedom. The temperature controlled RIE method is free from directional freedom but the quality of the mirror depends on process and materials. The reflow method is also subjected to lower throughput. The machining provides good surface profile; however, it is difficult to cut an individual waveguide on a substrate due to the physical size of the machining tool. We developed a new fabrication method using a microtome blade.

5.2.1   Coupling Efficiency Calculation

The coupling efficiency is one of the most critical issues in the fully embedded optical interconnects because of the concerns about thermal management and cross-talk. Higher coupling efficiency between a waveguide and VCSEL (or detector) enables the lower power operation of VCSEL. Furthermore, when small aperture VCSEL is used to operate at a high speed, for example, 3 μm aperture for 10 GHz operation, the coupling efficiency is a paramount concern because of the large spatial divergence of VCSEL's light. A large aperture selectively oxidized VCSEL operates in multiple transverse modes due to the strong index confinement created by oxide layer with low-refractive index [45]. Various techniques were introduced to operate in a single transverse mode [46–48]. Real spatial distribution of the VCSEL is not the same as Gaussian profile; however, I can consider it as the Gaussian profile by ignoring small discrepancies. This assumption results in a simple calculation. Another assumption is that lights within the acceptance angle of a waveguide are totally coupled into the waveguide. There are about ten supporting modes in the 50 μm square waveguide with $\Delta n = 0.01$. For an exact calculation, I have to consider all the modes, but the number of mode is quite large. It can be treated as geometrical optics.

To calculate coupling efficiency, I need to know intensity distribution and propagation angle at 45° mirror facet. Figure 5.1(a) illustrates a waveguide with a 45° mirror coupler and a

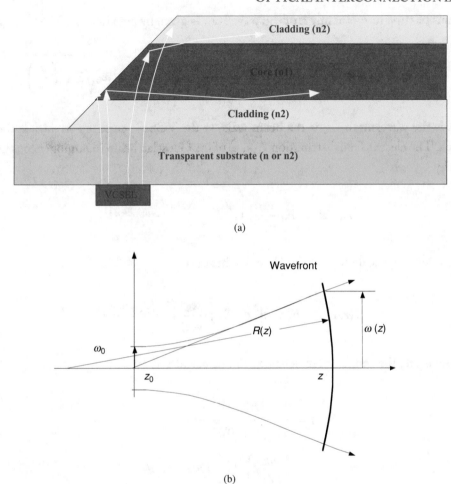

FIGURE 5.1: Coupling mechanism. Diagram of the coupling mechanism (a) and Gaussian beam propagation in a homogeneous medium (b)

transparent substrate. The VCSEL is bonded to the substrate; hence, laser light travels through the substrate and bends at the right angle at the mirror facet. The transparent substrate is optically isotropic. Figure 5.1(b) illustrates the Gaussian beam propagation in homogeneous medium.

The propagation angle at a mirror surface θ (r, z) can be calculated from the radius of curvature of wavefront $R(z)$ and distance from center r [49].

$$\theta(r, z) = \tan^{-1}\left(\frac{R(z)}{r}\right) \qquad (5.1)$$

The radius of curvature R at any z of the wavefront is given by equation

$$R(z) = z\left(1 + \frac{\pi \omega_0^2 n}{\lambda z}\right), \qquad z_0 = \frac{\pi \omega_0^2 n}{\lambda}, \qquad \omega(z) = \omega_0\sqrt{1 + \left(\frac{z}{z_0}\right)} \qquad (5.2)$$

where λ is the wavelength, ω_0 is the beam waist at the VCSEL surface and $\omega(z)$ is the beam width at z. The electric field distribution $E(r, z)$ of the Gaussian beam in homogeneous medium is given by

$$E(r, z) = E_0 \frac{\omega_0}{\omega(z)} e^{\left\{-i[kz-\delta(z)]-(r^2)\left[\frac{1}{\omega^2(z)} - \frac{ik}{2R(z)}\right]\right\}} \qquad (5.3)$$

And the intensity distribution of the Gaussian beam is

$$I(r, z) = |E(r, z)|^2 = E_0 \left\{\frac{\omega_0}{\omega(z)}\right\}^2 e^{\left(-2(r^2)\frac{1}{\omega^2(z)}\right)} \qquad (5.4)$$

Therefore, the coupling efficiency, η, can be calculated by

$$\eta = \frac{\int_{-r_c}^{r_c} |E(r, z)|^2 \, dr}{\int_0^\infty |E(r, 0)|^2 \, dr}$$

$$= \left(\frac{\omega_0}{\omega(z)}\right)^2 \int_{-r_c}^{r_c} |E(r, z)|^2 \, dr \qquad (5.5)$$

where, r_c is the maximum radius at the mirror facet which correspond to the acceptance angle of the waveguide.

The coupling efficiencies between VCSEL and a square (50 μm × 50 μm) waveguide with $\Delta n = 0.01$ (refractive index difference between a core and a cladding) were calculated as a function of angular deviation from $45°$. The substrate thickness (bottom cladding) and the aperture of the VCSEL are 127 μm and 12 μm, respectively.

Figure 5.2(a) shows the intensity distributions of laser light at the mirror surface, and the coupling efficiencies as a function of angular deviation from $45°$ for 127 μm thick substrate, 50 μm × 50 μm waveguide, and VCSEL with 12 μm aperture. The facet of $45°$ mirror was coated with aluminum to ensure reflection because TIR (total internal reflection) does not occur due to the top cladding layer. The reflectance of aluminum is about 92%. In this scheme all laser lights fall within the mirror. The coupling efficiency is 92% which means nearly 100% of the

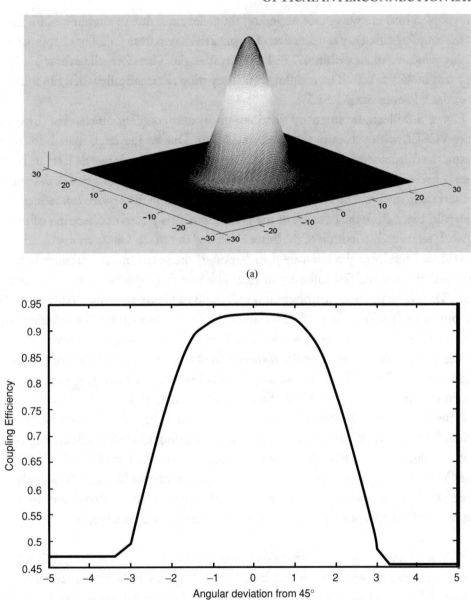

(a)

(b)

FIGURE 5.2: Intensity distributions at the mirror surface for 127 μm thick cladding (a) and coupling efficiencies as a function of angular deviation from 45° (b)

light is coupled into the waveguide excluding the reflectance due to aluminium. Figure 5.2(b) shows the coupling efficiency as a function of angular deviation from 45°. The coupling efficiency maintains constant values within 45° ± 1.5° mirror angle. Therefore, the mirror angle should be kept within 45° ± 1.5°. The coupling efficiency drops dramatically when the mirror angle is out of the tolerance range (±1.5°).

Figure 5.3 illustrates intensity distribution (a) and coupling efficiencies (b) for 3 μm aperture VCSEL with substrate thickness of 37 μm. Due to the large spatial divergence of the beam, the diameter of the beam spot increases more rapidly as it travels farther; hence, the distance between VCSEL and the mirror is kept short. As a result, the spot size of the beam at the mirror surface is smaller than the mirror itself. However, large divergence can reduce coupling efficiency. In Fig. 5.3(a), the position of the peak intensity is placed at off-center of the mirror. The global intensity distribution of the beam is skewed out of the Gaussian distribution caused from combined results of the intensity distribution of the beam and the distance between the VCSEL and the mirror. The intensity of light is inversely proportional to the square of the distance. Transverse intensity distribution is reduced by a Gaussian profile. Although all lights hit the mirror surface, only some of them coupled into the waveguide. The resultant coupling efficiency is 34% at 45°. However, for 40° tilted angle, the coupling efficiency is increased to 45.4% due to the offset of the intensity distribution at the mirror. The interesting point is that the minimum coupling efficiency is at the 45° tilted mirror. By increasing or decreasing the angle of mirror from 45°, the coupling efficiency will be increased.

If the distance between the 45° waveguide mirror and the VCSEL increases, some of the laser light hits the out of mirror surface; therefore, coupling efficiency will be dropped. This kind of situation occurs when a light source is mounted on the top of the PCB. For a small aperture VCSEL, the 45° mirror is not a good coupler as shown in Fig. 5.3. To enhance the coupling efficiency, a collimator, different types of coupler such as curved surface 45° tilted mirror, or a vertical optical via incorporated with 45° mirror will be required.

5.2.2   Fabrication of the 45° Micromirror Coupler

Polymers can cut with a very sharp blade. Based on this fact, a simple fabricating technique was developed. A blade sliding down to the waveguide substrate at 45° slope cuts the waveguides at 45°. It is like a guillotine sliding on a slope. The difference is that the blade of the guillotine falls at the right angle, on the other hand, the blade falls at 45°.

Two kinds of blades were tested. One is the typical razor blade, and the other is microma-chined silicon blade. Because the typical blade is very sharp, it can make a very smooth surface; however, a dual edged blade makes it difficult to cut at an exact angle of 45°.

The micromachined Si-blade also has an extremely sharp edge. The major fabrication process is the anisotropic chemical wet etch of a silicon wafer [Fig. 5.4]. Silicon nitride film with

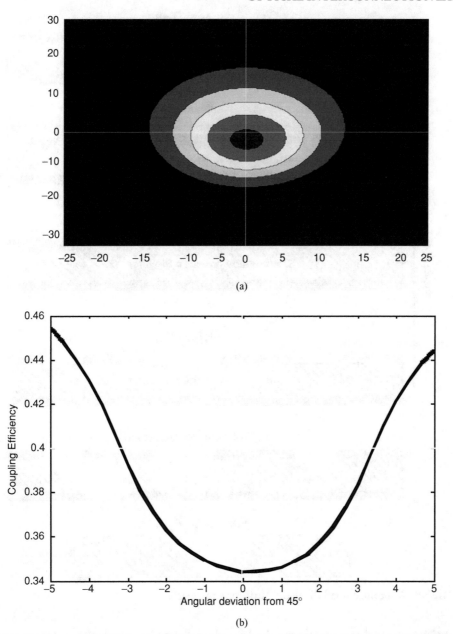

(a)

(b)

FIGURE 5.3: Intensity profile at the mirror surface (a) and coupling efficiencies (b). [VCSEL aperture: 3 μm, cladding thickness: 37 μm]

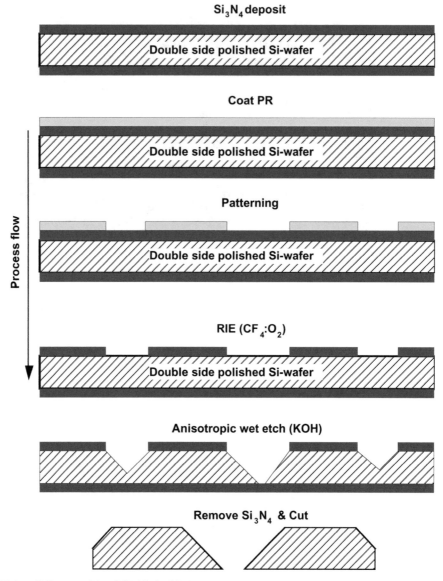

FIGURE 5.4: Micromachined Si-blade fabrication process

1000 Å thickness is deposited on the double side polished Si-wafer with $< 100 >$ orientation in a low-pressure chemical vapor deposition (LPCVD) chamber. Then, photoresist coating and patterning steps are followed. The next step is patterning Si_3N_4 film. As a reactive gas, tetrachlorocarbon (CF_4) and oxygen (O_2) mixture was introduced in a RIE chamber. The operating conditions of the RIE were 100 W RF power with the flow rate of 65 sccm and 38 sccm for CF_4 and O_2, respectively.

The following process is the anisotropic wet etching. A basic feature of anisotropic etchant is that their etch rates are strongly dependent on crystallographic orientation. Anisotropic etching is a function of areal density of atoms, the energy needed to remove an atom from the surface and geometrical screening effect [50]. The $< 111 >$ planes have an inclination of $54.74°$. When the Si-wafer is exposed to anisotropic etchant, etching stops at $< 111 >$ planes. The potassium hydroxide (KOH) solution of 40% was used as the anisotropic etchant. The chemical reaction equations are [51]

$$Si + 2OH^- \rightarrow Si(OH)_2^{2+} + 4e^- \tag{5.6}$$

$$4H_2O + 4e^- \rightarrow 4OH^- + 2H_2 \tag{5.7}$$

$$Si(OH)_2^{2+} + 4OH^- \rightarrow SiO_2(OH)_2^{2-} + 2H_2O \tag{5.8}$$

The overall reaction is summarized as

$$Si + 2OH^- + 2H_2O \rightarrow SiO_2(OH)_2^{2-} + 2H_2 \tag{5.9}$$

Final fabrication step is removing Si_3N_4 on the bottom side of the Si-wafer and dicing each blade.

The silicone blade has an apex angle of $54.74°$. When the blade tilt at $45°$, an angle between the waveguide substrate and the cutting plane of the blade exceeds $90°$ [Fig. 5.5]. The shear strain results from the obtuse angle. As a result of large shear strain, it is more like tearing than cutting. It caused a poor quality cutting plane. Microtome blade has an acute angle and sharp cutting edge; however, it is dual edged. The photograph of two blades is shown in Fig. 5.6. The apex angle of microtome blade and micromachined Si-blade are $18.3°$ and $54.7°$, respectively.

The material of the master waveguide structure is SU-8 (MicroChem) photoresist. The photoresist (SU-8) was spun on silicon wafer and then developed. Detailed process will be explained in the following section. The fabrication of $45°$ waveguide mirror was fabricated by tilted microtome setup [Fig. 5.7]. The master waveguide structure was kept at $120°C$ on a hot plate. In general, elevated temperature soften polymer. The softness ends up with smoother cutting surface. The blade was sliding down on substrate at $45°$ slope. The top-off view and surface of the mirror is shown in the Fig. 5.8. All waveguides were cut simultaneously by the microtome blade.

5.3 SOFT MOLDING

5.3.1 Introduction

There are various fabricating techniques to define optical waveguide on myriads of substrates. The reactive ion etch (RIE) uses ionized gas to remove material where it is not protected

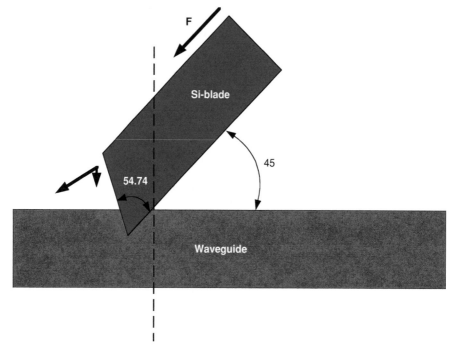

FIGURE 5.5: Shearing effect by micromachined Si-blade during cutting

by a mask material in a vacuum chamber. The size of the substrate purely depends on the vacuum chamber. It is relatively free from material selection because RIE is a physical removing process. The lithography uses optically transparent and photosensitive materials. The exposed or unexposed area by UV light makes the material insoluble to solvent due to the cross-linking of molecule. However, there is a limitation for choosing material due to the lack of materials which have optical transparency in the interesting region and photosensitivity. Hot embossing and molding are indirect fabrication techniques by means of transferring the waveguide structure on the substrate. Embossing plate or cast is first fabricated using the master waveguide pattern. Once the plate or the cast was fabricated, the rest of processes are purely replication steps. Therefore, these fabrication techniques are suitable for mass production like stamping of a compact disk. Laser ablation technique is similar to carving without using a chisel. The high-intensity UV laser beam removes the material of unwanted region. The motion stage which holds the waveguide substrate is moved along the predefined paths. Therefore, processing time is quite long. However, it is a quite versatile tool for small quantities in fabrication and does not require a mask pattern.

The molding method was chosen in this experiment because of its dependable process and suitability for large-volume production even though only a small quantity is needed in a

(a)   (b)

(c)   (d)

FIGURE 5.6: SEM pictures of micromachined silicon blade (a) and enlarged view of the blade edge (b), steel microtome blade (c), and enlarged view of the blade edge (d)

FIGURE 5.7: 45° tilted microtome apparatus setup

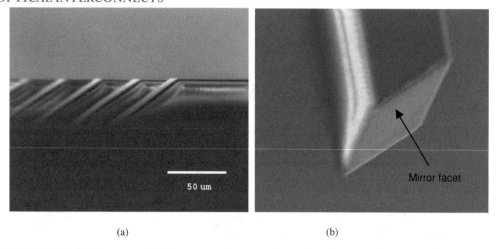

(a)   (b)

FIGURE 5.8: SEM photograph of the waveguide structures with 45° waveguide mirrors (a) and enlarged view of the mirror surface (b)

research stage. Solid mold is generally used in various applications such as embossing, optical disk stamping, and Fresnel lens fabrication. The solid mold is made of nickel alloy by electroplating. The fabrication of the solid mold has higher cost and takes a long time. These reasons make us to seek alternative mold materials. Curable resins such as silicone and urethane can be used as an alternative to reduce the fabrication cost and time. The soft mold has been used in various applications such as rubber stamp, small quantity manufacturing, replication, and micromachining. Specially, when the soft mold is used in microprinting and micromachining, the whole process is usually called as soft lithography.

5.3.2   Master and Mold Fabrication

The mold is a negative copy of the master structure. Once the master is fabricated, making mold is simply pouring a mold material over the master and curing. A silicone elastomer, especially poly(dimethylsiloxane) (PDMS) was chosen to fabricate the mold. PDMS has several unique characteristics. It has a low glass-transition temperature, stability at wide range of temperatures $(-50-200°C)$, resistance to most chemicals, low-interfacial free energy (21.6 dyne/cm), and good thermal stability [52, 53].

Master waveguide structure and mold fabrication process is shown in Fig. 5.9. The master for the mold was fabricated on a Si-wafer. Multimode waveguide is required for board level interconnection because of the requirement of the lower packaging cost. Alignment of the devices (laser and detector) and the waveguide is easier when the core size of the waveguide is large.

Designed size of the multimode waveguide is 50 μm × 50 μm. The process is the same as the standard photolithography. The Piranha bath consists of two parts of sulfuric acid (H_2SO_4)

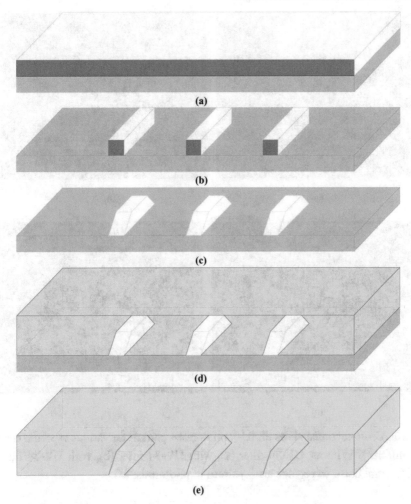

FIGURE 5.9: Master waveguide and mold fabrication process, spin coat photoresist (Su-8) at 400 r/min for 5 sec + 1500 r/min for 40 sec (a) and patterning (exposure, postexposure bake, and develop) (b), 45° cut by microtome (c), pouring PDMS over the master (d), and curing PDMS and remove the master (e)

and one part of hydrogen peroxide (H_2O_2). A Si-wafer was cleaned in the Piranha bath. After cleaning, the wafer was baked at 150°C to remove adsorbed water just before spin coating. This baking step improves adhesion and removes bubbles in prebaking step. After baking, photoresist (SU8-2000, MicroChem) was poured on the wafer and then spin coated at 400 r/min for 5 s and then ramped to 1500 r/min for 40 s. The wafer was kept on a leveled surface for 5 min to improve uniformity. And then, the wafer was moved to a leveled hot-plate for prebake. The pre-bake was carried out at 65°C for 5 min and at 90°C for 40 min.

(a)   (b)

(c)   (d)

FIGURE 5.10: Cross-sectional views of SU-8 photoresist pattern for various exposure conditions (Exposure: 300 mJ/cm^2). Without UV-34 filter (a), with UV-34 filter (b), with UV-34 filter and Index matching oil (c), and linear waveguide array pattern of photoresist (d)

The patterning is the trickiest part during the whole fabrication process. The SU-8 is chemically enhanced photopolymer; hence, postexposure bake is necessary. The photoresist is a negative type, i.e., unexposed area will be removed during the developing process. Exposure, to pattern the photoresist, is carried out using Carl–Suss aligner. Figure 5.10 shows cross-sections of SU-8 photoresist pattern for various exposure conditions. The Cross-linking of the SU-8 proceeds in two steps. The first step is the formation of acid during exposure. The second step is an acid initiated thermally driven epoxy cross-linking. The photoresist tends to have a negative sloped sidewall, which is not good for mold application. The sidewall should have a positive slope or at least be vertical for mold application. The exaggerated negative wall is often called as T-topping. The T-topping results from the lateral diffusion of the acid near the surface. The UV lights shorter than 350 nm are absorbed strongly at the top surface of the photoresist; hence, acid is generated by UV lights, which diffuses laterally along the top surface.

After the UV exposure, postexposure bake (PEB) process was carried out under the same conditions as described in the prebake step. If the PEB time is short, the pattern does not have enough adhesion; hence, the pattern is peeled off from the substrate during the developing process.

The T-toping can be removed by filtering out short wavelength below 350 nm [54]. The T-topping of the pattern is shown in Fig. 5.10(a). Nearly vertical sidewall [Fig. 5.10(b)] was made using short wavelength cut filter (UV-34, Hoya). However, there is still a beaked feature between sidewalls and a top surface. A beaked feature results from the diffraction at the interface between the mask and the photoresist, and it can be eliminated by filling index matching oil (glycerol) into the gap [55]. The ethylene glycol was used to fill the air gap in this experiment instead of using glycerol.

The beak was completely removed [Fig. 5.10(c)]. The portion of the 1 × 12 linear array of waveguide pattern is shown in Fig. 5.10(d). The final step in the master fabrication process is cutting waveguide ends at 45°. The cutting was completed by using the microtome setup as described in the Section 5.5.2.

The mold material is PDMS (Sylgard 184, Dow Corning). Prepolymer and curing agent were mixed at 1:10 ratio. Air bubbles trapped in PDMS were removed in a vacuum chamber. After removing air bubbles, PDMS was poured on the master and cured at 90°C in a vacuum chamber for 10 h. Surface relief structures were transferred from the master to the mold. The PDMS mold is shown in Fig. 5.11.

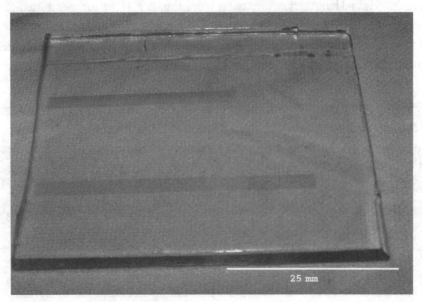

25 mm

FIGURE 5.11: PDMS mold

5.3.3 Deformation Compensation

Materials used in the waveguide fabrication process have different coefficients of thermal expansion (CTE). For example, the COC (cycloolefin copolymer, Topas 5013 from Ticona) as a cladding material has the CTE of 60 ppm/K. The CTEs are 2.3, 52, 30 ppm/K for silicon, SU-8, and PDMS, respectively. The dimension of the cast is not the same as the master waveguide due to the shrinkage of PDMS cast after cooling down to ambient temperature. There are two approaches to compensate for dimensional change. One is making a compensated mold after trial and error and/or accurate calculation. The other is the compensation technique during a molding process by adjusting mold pressure. This approach is only valid for a soft cast. The first approach is suitable for fixed materials and well-defined process conditions. If a material or process is changed, the cast also must be changed. The compensation technique during the mold process was chosen because the cast material is soft, and it is easy to adjust process in a research stage.

Dimensional change occurs at the beginning of the process. The first step to make the flexible waveguide film starts with the fabrication of the soft cast. To make a cast, curing procedure at certain temperature is followed after pouring liquid silicone elastomer on to the master structure. During this process, dimensional change occurs in both the master and the mold. The dimensional change of the master at elevated temperature is given by Eq. (5.10).

$$L = L_O(1 + \alpha_{Si} \bullet \Delta T) \qquad (5.10)$$

where L_O is the original length at reference temperature, ΔT is the temperature difference, and α_{Si} is the linear expansion coefficient of silicon. The length of the mold at curing temperature is equal to that of the master structure. After cooling down the mold to ambient temperature, the mold is shrunken. The length after cooling (L_{cast}) can be calculated from the linear expansion

TABLE 5.1: Physical Properties of Materials Used in Molding Process.

MATERIAL	COEFFICIENT OF THERMAL EXPANSION, α (PPM/$^\circ$K)	YOUNG'S MODULUS, E (MPA)	POISSON'S RATIO, v
Silicon	2.33	202,000	0.27
SU-8	52	4020	0.22
PDMS	30	0.75	0.5
Topas5013	60	–	–

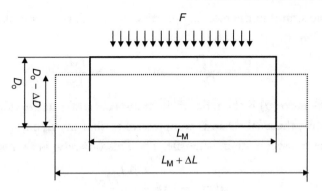

FIGURE 5.12: Mold deformation during compression molding

equation [Eq. (5.11)].

$$L_{cast} = L_O(1 + \alpha_{Si} \bullet \Delta T)(1 - \alpha_{PDMS} \bullet \Delta T)$$
$$= L_O(1 - \alpha_{PDMS} \bullet \Delta T + \alpha_{Si} \bullet \Delta T - \alpha_{Si} \bullet \alpha_{PDMS} \bullet \Delta T^2)$$
$$\cong L_O(1 - (\alpha_{PDMS} - \alpha_{Si})\Delta T) \tag{5.11}$$

The dimensional change of the final waveguide film occurs during the molding process. The mold is made of elastic rubber; hence, the applied pressure can deform the mold. In addition, the molding process accompanies pressure and heat. The waveguide substrate and the mold expand due to the thermal expansion. The length of the waveguide substrate (Topas film) (L_{Topas}) and the length of the mold, due to the thermal expansion are,

$$L_{Topas} = L_O(1 + \alpha_{Topas} \bullet \Delta T_M) \tag{5.12}$$

$$L_{cast, M, therm} = L_O(1 - (\alpha_{PDMS} - \alpha_{Si})\Delta T)(1 + \alpha_{PDMS} \bullet \Delta T_M) \tag{5.13}$$

If mold curing temperature (ΔT) is the same as the molding temperature (ΔT_M)then the Eq. (5.13) is simply,

$$L_{cast, M, therm} = L_O(1 + \alpha_{Si} \bullet \Delta T) \tag{5.14}$$

The dimension of the cast is also affected by the applied pressure (F) during the mold process.

When the cast is subjected to a load, it deforms not only in the direction of the load, but also in the direction perpendicular to the load. In the molding process, the load is compressive. The length in the direction of the pressure will decrease, and transverse length will increase.

There are two strains in the molding process, one is axial (ε_a), and the other is transverse (ε_t) [56].

$$\varepsilon_a = \frac{\Delta D}{D_O} \quad \text{and} \quad \varepsilon_t = \frac{\Delta L}{L_M} \tag{5.15}$$

The Poisson's ratio (υ) is the ratio of the transverse strain to the axial strain. If pressure is applied on the cast, the axial strain is compressive so that the sign of strain is negative. On the other hand, the transverse strain is tensile. The Poisson's ratio is always a positive value.

$$\upsilon = -\frac{\varepsilon_t}{\varepsilon_a} = -\frac{\Delta L/L_M}{\Delta D/D_O} \tag{5.16}$$

The length change of the cast due to the pressure can be calculated by introducing Young's modulus (E).

$$E = \sigma\frac{D_O}{\Delta D} \tag{5.17}$$

Here, σ is the applied load (N/m^2).

The final length change of the cast accounting thermal expansion and deformation [Eq. (5.18)] can be calculated by plugging Young's modulus term into the Poisson's equation.

$$\Delta L = -L_{\text{cast, M, therm}}\frac{\upsilon\sigma}{E} \tag{5.18}$$

The displacement (ΔL) of the mold should be equal to the displacement of the waveguide film.

$$-L_M\frac{\upsilon\sigma}{E} = L_O\alpha_{\text{Topas}}\Delta T \tag{5.19}$$

Therefore, optimized pressure (σ) to compensate for deformation is given by Eq. (5.20).

$$\sigma = -\frac{E\alpha_{\text{Topas}}\Delta T}{(1 + \alpha_{\text{Si}}\Delta T)\upsilon} \tag{5.20}$$

Data in Table 5.1 were used during the calculation. The optimized molding pressure was 6000 Pa. If the molding pressure is less than 6000 Pa, the distance between two waveguides will be shorter than the designed distance. In case of the pressure greater than 6000 Pa, the distance will be larger. Total displacement will be accumulated by increasing mold or feature dimension.

5.3.4   Optical Interconnection Layer Fabrication

The fully embedded board level optical interconnection requires a thin flexible optical layer. Current electroplating technology can easily plate a through-hole or a via having an aspect ratio of one in production line and can plate a hole having an aspect ratio of three in laboratory. The

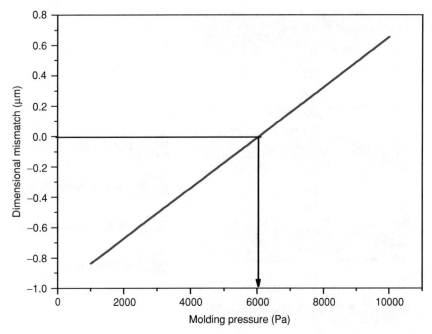

FIGURE 5.13: Dimensional change as a function of molding pressure. [Dimensional change is based on 250 μm structure, i.e., amount of deformation will be accumulated if the size of a feature or mold increased]

size of a typical electrical pad on the device is about 100 μm. These are the main reasons for the thickness limit of substrate film. The thin and flexible optical waveguide layer was fabricated by compression molding technique using soft mold. A 127 μm thick optically transparent film (Topas 5013) was used as a substrate of the waveguide circuit.

The fabrication step is straightforward. First, the core material (SU-8) was poured on the heated PDMS, which was kept at 50°C [Fig. 5.14(a)]. The heated PDMS mold suppressed bubble generation during the molding process. And then, excess SU-8 was scraped out using squeegee [Fig. 5.14(b)]. The squeegee also was made of PDMS. The Topas film was applied on the top of the PDMS mold filled with SU-8. In the next step, the mold and the Topas film were inserted into the press machine and then was applied with the pressure of 6000 Pa for 30 min while plunge plate was held at 90°C [Fig. 5.14(c)]. The cooling down procedure was followed. In this procedure, the mold pressure decreased gradually due to the thermal contraction. The core material (SU-8) was transferred to the substrate film [Fig.5.14(d)]. In the next step, the substrate film without the top cladding was exposed to UV lights to cross-link the SU-8. Once the film was exposed, it becomes chemically and thermally stable. Aluminum was deposited on the mirror facets in a vacuum chamber to make the mirror. Finally, top cladding material

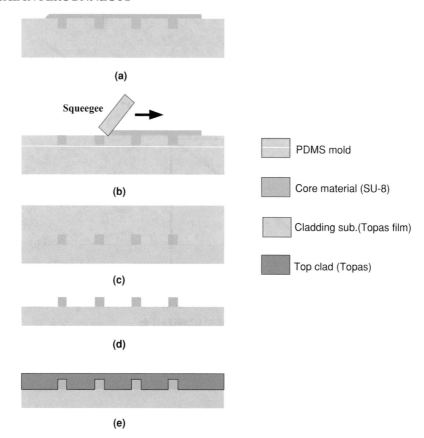

FIGURE 5.14: Optical interconnection layer fabrication process flow. Apply core material (SU-8) (a), remove excess material (b), apply heat and pressure on the mold and cladding substrate (c), cool down and release plunger (d), and apply top cladding materials (either spin coat or lamination) (e)

(Topas) was coated on the film. Fabricated optical interconnection layer is shown in Fig. 5.15. It has micromirror couplers and 12 channel waveguides of 50 mm in length.

5.4   PROPAGATION LOSS MEASUREMENT

The absorption loss of Topas film was determined by the transmittance of material using spectrometer. Measured transmittance values according to wavelengths were compared to the theoretical values.

The transmittance of medium is the ratio of the sum of all transmitted lights to the incident power. α is the absorption coefficient of the material, L is the thickness of the medium, and R is reflectance between interfaces [Fig. 5.16].

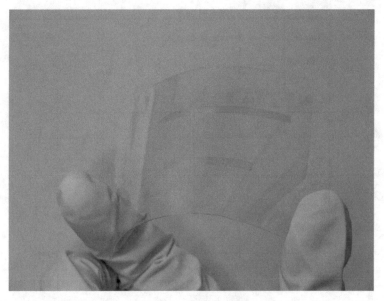

FIGURE 5.15: Fabricated flexible optical interconnection layer

For normal incident case and air-to-medium interface, the reflectance at the interface is given by Eq. (5.21) [14].

$$R = \left(\frac{n-1}{n+1}\right)^2 \qquad (5.21)$$

Here, n is the refractive index of the medium.

The refractive index of Topas 5013 is calculated from Cauchy equation [Eq. (5.22)].

$$n = A + \frac{B}{\lambda^2} + \frac{C}{\lambda^4} \qquad (5.22)$$

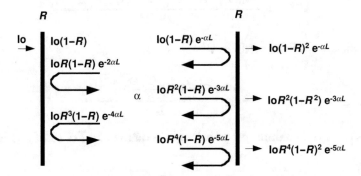

FIGURE 5.16: Reflectance and transmittance of thick medium

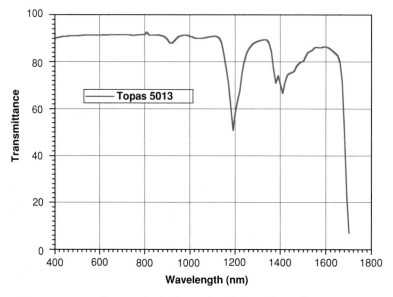

FIGURE 5.17: Transmittance of 3 mm thick Topas 5013 [from Ticona]

For Topas 5013 medium, the coefficients are $A = 1.519808$, $B = 440569.2$, and $C = 7.755498 \times 10^{11}$ [from Ticona]. The transmitted intensity is the sum of all the transmitted lights by multiple reflections at two interfaces [Figs. 5.17 and 5.18].

$$T = (1 - R)^2\, e^{-\alpha L} \left[1 + R^2\, e^{-2\alpha L} + R^4\, e^{-4\alpha L} + R^6\, e^{-6\alpha L} + \cdots \right]$$

$$= (1 - R)^2\, e^{-\alpha L} \left[\frac{1}{1 - R^2\, e^{-2\alpha L}} \right] = \frac{(1 - R)^2\, e^{-\alpha L}}{1 - R^2\, e^{-2\alpha L}} \qquad (5.23)$$

$$e^{-\alpha L} = \frac{-I_0(1 - R)^2 \pm \sqrt{(1 - R)^4 + 4T^2 R^2}}{2 T R^2} \qquad (5.24)$$

Absorption coefficient α of the medium is

$$\alpha = -\frac{\ln\left(\frac{-I_0(1-R)^2 \pm \sqrt{(1-R)^4 + 4T^2 R^2}}{2 T R^2} \right)}{L} \qquad (5.25)$$

Absorption loss α_a of the medium per unit length is defined by Eq. (5.26). [Fig. 5.19].

$$\alpha_a = \frac{10}{l} \mathrm{Log}_{10} \frac{P_{\mathrm{in}}}{P_{\mathrm{out}}} \qquad (5.26)$$

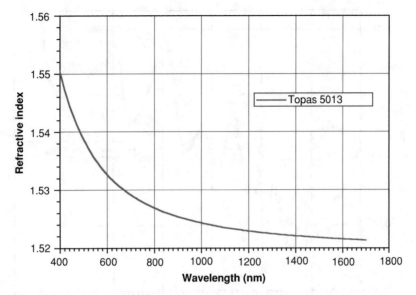

FIGURE 5.18: Refractive index of Topas 5013 as a function of wavelength. [from Ticona]

Here, P_{in}/P_{out} is the ratio of the input power to the output power after propagating medium and equals to $e^{\alpha L}$, then

$$\alpha_a = \frac{10}{L}\mathrm{Log}_{10}\, e^{\alpha L} \tag{5.27}$$

The refractive index of UV-exposed SU-8 as a function of wavelength was measured using transmittance curve. The oscillating curve beyond 650 nm wavelength did not represent real refractive index. There was some interference due to the thin sample. The refractive index is shown in Fig. 5.20. The refractive index was 1.584 at 850 nm wavelength.

The absorption loss of the Topas is 0.01 dB/cm and 0.03 dB/cm at 630 nm and 850 nm, respectively. The minimum absorption loss at 790 nm does not represent an exact value because there is a change of grating during wavelength scan.

Curtis reported 6 dB/cm propagation loss at 850 nm for a multimode waveguide [57]. Wong *et al..* reported low propagation loss of 0.22 and 0.48 dB/cm at 1330 and 1550 nm, respectively, using electron beam direct writing [58].

Waveguide propagation loss was measured by cut back method. The core dimension was 50 μm $\times$ 50 μm. The bottom cladding of the waveguide is 3 μm thick SiO_2. Fiber pig-tailed 850 nm laser was used to couple the laser to waveguide. The diameter of fiber is 10 μm, which is similar to the VCSEL aperture. Coupled out powers according to the length is shown in Fig. 5.21. The measured propagation loss was 0.6 dB/cm at 850 nm wavelength.

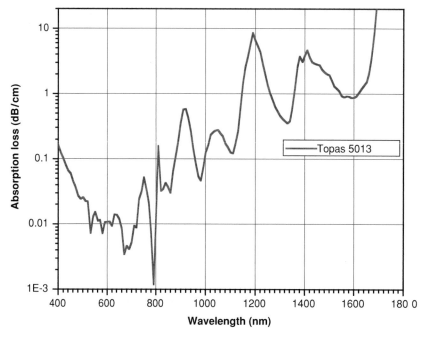

FIGURE 5.19: Extracted absorption loss of the Topas 5013 as a function of wavelength

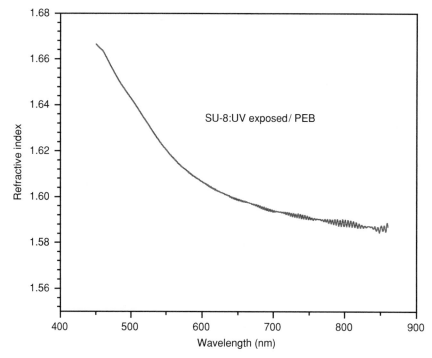

FIGURE 5.20: Measured refractive index as a function of wavelength

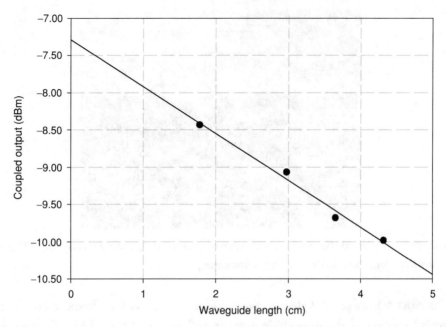

FIGURE 5.21: Coupled out power as a function of waveguide length. [Cut-back method, slope of the line is the propagation loss, the propagation loss is 0.6 dB/cm at 850 nm wavelength]

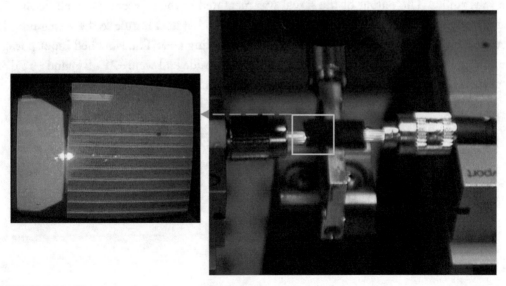

FIGURE 5.22: Test setup for the cross-talk measurement

FIGURE 5.23: Coupled out beams from 45° waveguide mirrors. [Flexible waveguide, 633 nm He–Ne laser was lunched from the other side of the wave guide]

The cross-talk of parallel channel waveguides is an important factor in communication. The channel spacing of the waveguide array is 250 μm, and the width of waveguide is 50 μm. The difference of refractive indices between the core and the cladding is about 0.058. To measure cross-talk, the 5 cm long channel waveguide was put on an autoaligner. And then, the fiber coupled laser light with the wavelength of 630 nm was lunched into one channel among the waveguides. The output of the signal was measured at the adjacent channel. Figure 5.22 shows a test setup. Laser light was launched from the end of fiber ferrule to the waveguide. The picture, left side of Fig. 5.22, is enlarged view of coupling area. The launched input power at the waveguide and output power at the adjacent waveguide end were −21 dBm and −53 dBm, respectively. Measured cross-talk between adjacent channels was 32 dB.

CHAPTER 6

System Integration

6.1 INTRODUCTION

The hybrid integration of optical layer with electrical layers is the most important part in the realization of optical interconnects. The PCB technologies is already well matured; therefore, great deviation from the current fabrication technique may result in the failure of commercialization. Hence, we have to minimize the change of fabrication process. The optical interconnects in board level cannot replace all the copper lines. There are slow interconnecting lines such as data, control, power, and ground lines. Especially, slow data lines do not need to be replaced by optical means because the fabrication cost and process of copper lines are very cheap and reliable.

The high-integration cost must be avoided. This requirement can be satisfied by separating fabrication processes. Obviously, the fabrication of waveguide layer and the integration of optoelectronic devices must be cheap. The optical interconnection layer and the electrical layers are fabricated independently. At the final integration step, two different types of layers are laminated together using matured lamination process.

Interface between optoelectronic device and control device is important because of the requirement for high-density (HD) packing. Laser drilled micro-via technology is utilized to satisfy HD packing. The thermal management of VCSEL is also important because of reliability concerns. Embedded VCSEL cannot be replaced to repair in a fully embedded integration; therefore, we have to be aware of good VCSEL before the integration and provide optimal operating condition.

6.2 INTEGRATION OF THIN-FILM VCSEL AND DETECTOR ARRAYS ON OIL

The integration of optoelectronic devices with the flexible waveguide film is the most important process among the whole integration steps including the final laminating process with the PCB. The flexible waveguide film has the thickness of 127 μm. The VCSEL and photodetector arrays have the diameter of about 95 μm. In general, copper electroplating allows plating of a via with

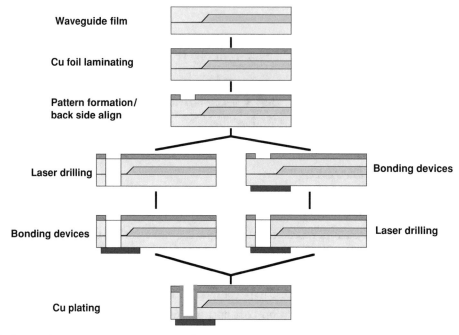

FIGURE 6.1: Devices integration process flow chart

aspect ratio of one; however, it is possible to plate a via with an aspect of three in a laboratory environment. The maximum diameter of the laser drilled via is limited by the pad size of devices, the aspect ratio, and the registration error during the lamination process. Figure 6.1 illustrates the device integration process.

First, one mil (25.4 μm) thick copper foil is laminated on the top of the flexible waveguide layer by applying heat and pressure. And then, this copper foil is patterned to form the top electrical pads for VCSEL and photodetector. The main reason for the formation of the laminated copper foil is the limitation of electroplating. The thickness of additional electrical layers easily exceeds 1 mm, and the diameter of the device pad is 95 μm. This translates to an aspect ratio of 100; therefore, this hole cannot be electroplated. The aspect ratio of via can be reduced by introducing the copper foil just above the waveguide layer; hence, we can electroplate micro-via. Furthermore, the patterns on the copper foils can be bigger. This means that larger registration error can be allowed during the lamination process with electrical layers.

The next step is either laser drilling or device bonding on the waveguide layer. The SEM pictures of laser drilled vias are shown in Figs. 6.2–6.4. Laser drilling and taking SEM picture were performed at Sanmina-SCI.

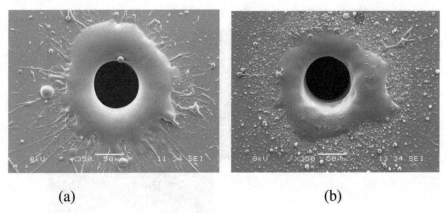

(a)   (b)

FIGURE 6.2: SEM pictures of the laser drilled micro-via. Frequency tripled Nd:YAG (wavelength: 355 nm), 127 μm thick Topas film. Incident side (a) and exit side (b). [from Sanmina-SCI]

Micro-via on the Topas film is shown in Fig. 6.2. Frequency tripled Nd:YAG laser, which has a wavelength of 355 nm, was used. The Topas film is relatively transparent at 355 nm wavelength. Melting and spattering around the hole were observed at both the incident and the exit side of the film. If copper foil was laminated on the top of Topas film, spatter and melt flow will be reduced. The reduced energy of laser pulse and short pulse width also help clean drilling. Another choice is using the UV-excimer laser which has shorter wavelength.

Micro-vias drilled with CO_2 laser are shown in Fig. 6.3. During the laser drilling, a device fell off from the Topas film. The drilling conditions were fixed pulse count and various pulse widths. When the Topas film was exposed to five series pulses of CO_2 laser which had 10 μs pulse width, only bump was created on the incident side. There was no clear hole when the pulse width of the laser was increased to 30 μs.

Another example of CO_2 laser drilled vias on the Topas film is shown in Fig. 6.4(a). The series of laser drilled vias in the upper portion of Fig. 6.4(a) were made by fixed laser pulse width of 10 μs and various pulse counts from 1 to 12. Via spacing was 250 μm. By increasing the pulse count, the melted region was expanded. While the pulse count was fixed, the pulse width increased from 1 to 11 μs, and the melted region, shown in the bottom portion of Fig. 6.4(a), was increased. The CO_2 laser cannot be used to drill via on the Topas film, which is the cladding material of the flexible waveguide film.

The picture of the UV-Nd:YAG laser drilled via on the waveguide film which is integrated with the device is shown in Fig. 6.4(b). The UV-Nd:YAG laser makes extensive damage on the film and it seems to damage the metal pad of the devices as well. Therefore, the integration of devices on the waveguide film should be done after the formation of micro-via drilling.

Incident side   Exit side

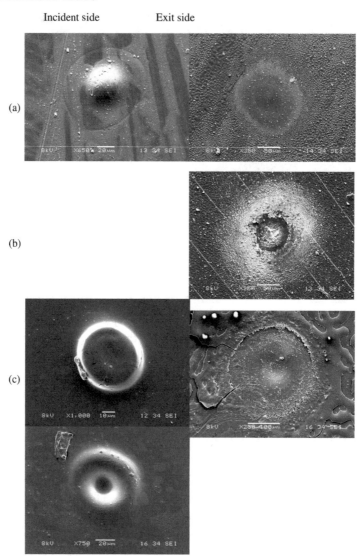

(a)

(b)

(c)

FIGURE 6.3: SEM pictures of the CO_2 laser drilled micro-vias. (wavelength: 10.6 μm), 127 μm thick Topas film. Five pulses with 10 μs pulse width (a) , five pulses with 20 μs pulse width (b), and five pulses with 30 μs pulse width (c). [from Sanmina-SCI]

The thin-film devices are usually bonded on an exotic substrate by Van der Waals force. Palladium is the most common material for bonding [59]. Although the bonding temperature is low, long bonding time is required. This limits the real world applications from the view point of throughput.

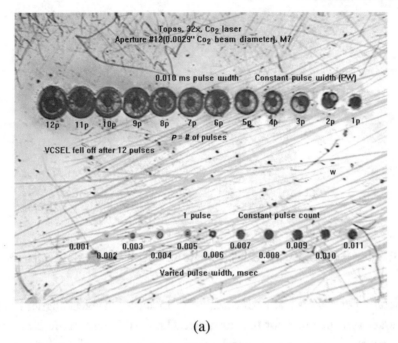

(a)

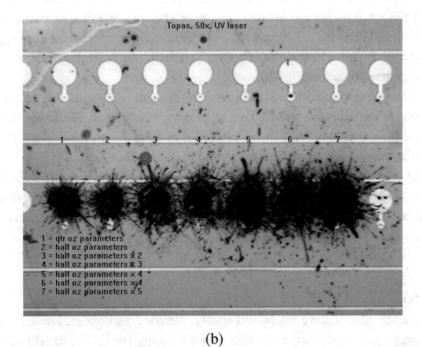

(b)

FIGURE 6.4: Pictures of laser drilled vias under various drilling conditions, CO_2 laser (a) and UV-Nd:YAG laser (Frequency tripled) (b). [from Sanmina-SCI]

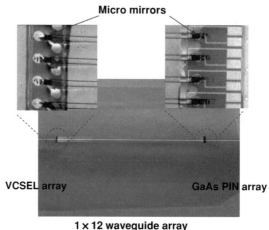

FIGURE 6.5: Integrated VCSEL and detector arrays on a flexible optical interconnection layer

The bonding of devices was performed using an aligner. Typical aligner has two holders; one for the mask and the other for the substrate. The flexible waveguide film was temporally bonded to a clear glass plate using water and placed on a mask holder. The device to be integrated was put on the substrate holder. A small amount of UV-curable adhesive was applied on the top of the device. When the device and the waveguide micromirror coupler were aligned, they were exposed to UV lights to cure the adhesive. The flexible optical interconnection layer with device integration is shown in Fig. 6.5.

The bonding of the device with the waveguide film can be accomplished by melt bonding without using UV-curable adhesive. When alignment is completed, the device is heated just above the melting temperature of the waveguide film for a short period. And, the device is bonded to the waveguide film without deforming the micro-vias.

So far, the integration of devices on the flexible waveguide film is accomplished. The final step is electroplating. The integrated waveguide film was submerged in copper electroplating solution plate to the sidewalls of micro-vias and device pad. Now, the optical interconnection layer is ready to be laminated with the electrical layers.

6.3   HYBRID INTEGRATION OF OIL AND PRINTED CIRCUIT BOARD

At the beginning of designing the hybrid system, electrical and optical layouts are designed simultaneously by incorporating each other. After finishing the layout, the electrical and the optical layer are fabricated separately. The electrical fabrication process can be a build-up process,

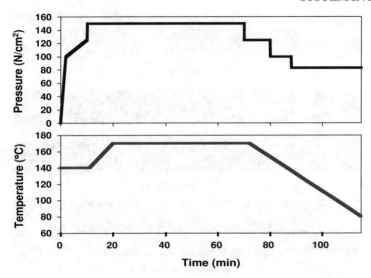

FIGURE 6.6: Industry standard FR-4 PCB laminating pressure and temperature. [Source: LLFa GmbH, Hanover/Germany]

a laminating process, or a mixed one. The fabricated optical interconnection layer will be inserted into the electrical layers. The industry standard PCB-lamination process uses high temperature and pressure to laminate layers. A typical laminating condition is shown in Fig. 6.5. The peak temperature and pressure reached at 170°C and 150 N/cm², respectively. The glass transition temperature of the Topas 5013 is only 135°C. However, the melting temperature of the waveguide core material, UV cross-linked SU-8, exceeds 200°C. Therefore, the optical interconnection layer cannot survive the standard laminating process. This is one of the reasons for building the electrical layers and the optical layer separately.

The hybrid-laminating process is shown in Fig. 6.6. The optical interconnection layer was already fabricated [Fig. 6.6(a)]. The electrical layers are partially laminated using the standard laminating process, i.e., upper electrical layers above the OIL and lower electrical layers below the OIL [Fig. 6.6(b)]. For better heat removal, the OIL is inserted near the bottom electrical layer. The shorter distance to heat sink provides better heat removing. In the next step, prepatterned bonding films, which have lower melting temperature than that of Topas, are applied on both sides of the OIL [Fig. 6.6(c)].

The following step is laminating OIL and electrical layers together by applying heat and pressure [Fig. 6.6(d)]. The final step is electroplating thick copper on the bottom of VCSEL and blind vias, which are used to establish electrical path between the top electrical pads and the OIL pads [Fig. 6.6(e)].

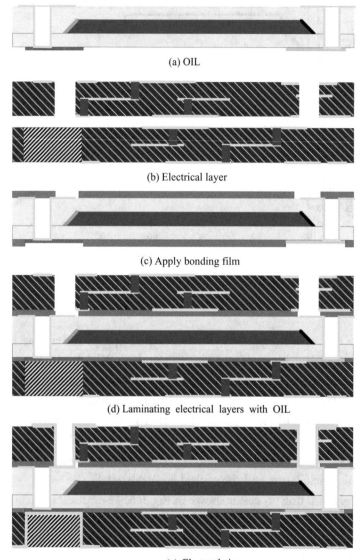

(a) OIL

(b) Electrical layer

(c) Apply bonding film

(d) Laminating electrical layers with OIL

(e) Electroplating

FIGURE 6.7: Flow chart of laminating electrical and optical interconnection layers

CHAPTER 7

Summary

This book describes a fully embedded board level optical interconnects in detail including the fabrication of the thin-film VCSEL array, its characterization, thermal management, the fabrication of optical interconnection layer, and the integration of mentioned devices on a flexible waveguide film.

All the optical components are buried within electrical PCB layers in a fully embedded board level optical interconnect. Therefore, we can save foot prints on the top real estate of the PCB and relieve packaging difficulty reduced by separating fabrication processes. To realize the fully embedded board level optical interconnection, very thin laser and photodetector are required. The signals to drive VCSEL and signals from photodetector flow through the electrical micro-vias.

An 1 × 12 array of 10 μm thick thin-film VCSEL, emitting at 850 nm wavelength, was fabricated by substrate removal technique. The GaAs substrate was completely removed by lapping and wet chemical etching. Measured L–I curve shows that 10 μm thick VCSEL did not show thermal rollover even at a high-injection current level. Measured thermal resistances of 10 μm thick VCSEL has the lowest value, nearly half of the 250 μm thick VCSEL. The thin-film VCSEL shows superior characteristics, higher external quantum efficiency, and lower thermal resistance over thicker VCSEL.

Thermal runaway of the embedded VCSEL is the critical concern due to the reliable operation of VCSEL for a long time. I investigated an effective heat sink structure for VCSEL through the 2D finite element thermal analysis. The 30 μm thick, directly electroplated, copper film on the back side of VCSEL array turns out to be an excellent heat sink without sacrificing the strategy of easy packaging. We also found the maximum thickness of VCSEL which guaranteed reliable operation. Extremely thin VCSEL is very difficult to handle. As it turns out, 70 μm is the maximum thickness for a reliable operation in a fully embedded structure. This thick device can be handled by a state-of-the-art automated pick and place machine.

The 45° waveguide micromirror couplers were fabricated by cutting waveguide ends at 45° using a microtome blade. The coupling efficiency of aluminum coated micromirror was calculated. The coupling efficiency between 12 μm aperture VCSEL mounted at the bottom of 127 μm thick cladding layer and 50 μm square waveguide was 92%, nearly 100%, ignoring the reflectance of aluminum. The coupling efficiency did not change within $\pm 1.5°$ angular deviation from $45°$. For a small aperture (3 μm) VCSEL and a thin bottom cladding layer (substrate), coupling efficiency was dropped to 34%. In this case, a typical $45°$ mirror cannot be used, and the other types of coupler, such as curved mirror or vertical optical via, are required to improve the coupling efficiency.

The optical interconnection layer was fabricated by compressive soft molding process. Molding is a suitable process for mass production. Mold was made of PDMS rubber. The micromirrors and waveguide were fabricated at the same time. Mold expansion during the process was compensated by adjusting the mold pressure. The core material of the waveguide was SU-8. The measured propagation loss of the waveguide was 0.6 dB/cm at 850 nm.

VCSEL and photodetector array were integrated on the flexible waveguide film. UV-Nd:YAG laser or CO_2 laser drilled micro-via was formed on the flexible waveguide film. The quality of micro-via is not good enough at this time. However, if an excimer laser is used to drill vias, the quality of via will be significantly improved.

The hybrid integration of the optical interconnection layer and electrical layers is ongoing at Sanmina-CSI. The reliability of the fully integrated system will be addressed in a future project.

CHAPTER 8

Effects of Thermal-Via Structures on Thin Film VCSELs for a Fully Embedded Board-Level Optical Interconnection System

8.1 INTRODUCTION

With rapid evolution of integrated circuit technology, the clock frequency and integration density of microprocessors improve each year [60]. However, the major bottleneck of high-speed interconnections on printed circuit boards (PCBs) is the limited data transmission rate of copper transmission lines with low-K material (FR-4). At a high-frequency operation (>10 GHz), copper transmission lines on a PCB provoke degradation in the rise and fall times of electrical signals, electromagnetic interference, and need higher power consumption due to skin effect and impedance mismatch [4, 61]. Various ideas on optical interconnection techniques are applied to overcome the frequency-dependent loss of electrical interconnection lines [61–63]. Although the optical interconnections have great advantages compared to the electrical interconnections, the reliability of these systems due to packaging vulnerability is a paramount concern.

To relieve packaging difficulty, we have developed a fully embedded board-level optical interconnection system, which is also depicted in Fig. 8.1 [41, 64–66]. All the optoelectronic components including vertical-cavity surface-emitting laser (VCSEL) array, PIN photodiode array, TIR coupling mirror, and planar polymer waveguides are integrated within the 3D interconnection layers during the conventional PCB fabrication processes. A 12-channel 850 nm GaAs VCSEL array is employed to convert electrical signals to optical signals. Optical signals are transmitted through a 12-channel polymer waveguide array and then converted to electrical signals by a 12-channel GaAs PIN photodiode array. Through the micro-via structures, electrical signals and bias are transmitted from an electrical layer to an embedded optical layer and vice versa.

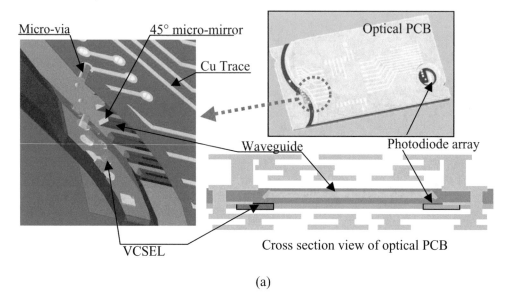

Micro-via   45° micro-mirror

Cu Trace

Optical PCB

Waveguide

Photodiode array

Cross section view of optical PCB

VCSEL

(a)

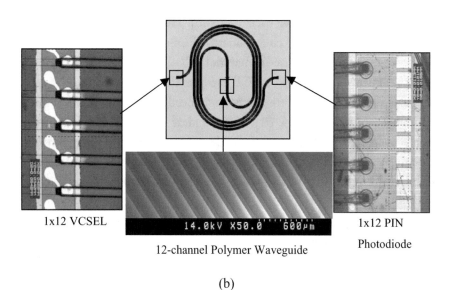

1x12 VCSEL

14.0kV X50.0 600μm

12-channel Polymer Waveguide

1x12 PIN Photodiode

(b)

FIGURE 8.1: Schematic diagrams of the fully embedded board-level optical interconnection system. (a) 3D cross-section view of an optical layer interposed PCB, (b) 12-channel optical waveguide film layer with integrated thin film VCSEL and PIN photodiode arrays

In this architecture, however, the self-heating effect of the VCSEL causes the critical issues in the system reliability since integrated VCSEL arrays are surrounded by thermal insulators such as optical polymer films (TOPAS®) and PCB bonding materials (prepreg or pressure-sensitive-adhesive film). Because the operating lifetime of the VCSEL decreases exponentially with temperature [67], the thermal management of the embedded VCSEL arrays is one of the prime concerns in the fully embedded optical interconnection system. Chen *et al.* [68] and Liu *et al.* [69] have reported device-level investigations on the thermal characteristics of the VCSEL. Also, comprehensive studies on the thermal resistance of an integrated VCSEL array on a PCB were performed by Lee *et al.* [31], Pu *et al.* [32], and Krishnamoorthy *et al.* [70]. Those papers presented valuable results, but the results are not applicable to our system directly because of the different integration structures.

This chapter presents theoretical and experimental studies of the thermal characteristics of the fully embedded thin film VCSEL array, which determines the effective thermal-via structures. The thermal resistances as a function of the substrate thickness of the VCSEL are experimentally measured. 2D finite-element analysis is performed to simulate the temperature field distribution near and across the active region inside the VCSEL as a function of the substrate thickness of the VCSEL and the thermal-via structure. Not only the heat generation in the active region but also the joule heating ($I^2 R$) effect in the distributed Bragg reflectors (DBRs) is considered by the thermal-electric direct coupled-field analysis.

8.2 FABRICATION OF A THIN FILM VCSEL AND MEASUREMENT OF THERMAL RESISTANCE (R_{th})

A 12-channel 850 nm oxide-confined VCSEL array with 10 Gbps each is employed as an input light source. The initial 200μm thick GaAs substrate of the VCSEL is removed not only for facilitating the fully embedded integration structure but also for managing the VCSEL temperature as previously mentioned. The initial GaAs substrate of the VCSEL is reduced down to 100 μm by mechanical lapping and polishing processes. After the mechanical thinning, the thickness of the VCSEL, within the range from 100μm to 10μm, is precisely controlled by the wet chemical etching process [71]. Figures 8.2(a) and (b) show a part of the VCSEL array before and after the substrate thinning processes, respectively. The CW L–I and I–V characteristics of 200 μm and 20 μm thick VCSELs are shown in Fig. 8.2(c). The measured threshold current and slope efficiency are 0.7 mA and 0.55 mW/mA, respectively. The measured characteristic temperature (T_o), indicating the temperature sensitivity of the threshold current, is in the range of 150–155 K for the studied VCSEL. Figure 8.3(a) shows the schematic diagram of the measurement setup for optical properties. Figure 8.3(b) shows a 10 Gbps eye-diagram of a 20 μm thick VCSEL measured by a digital communication analyzer (HP-83480A). Electrical contact is made with micro-probes linked to a pulse pattern generator (HP-83592C). The pulse

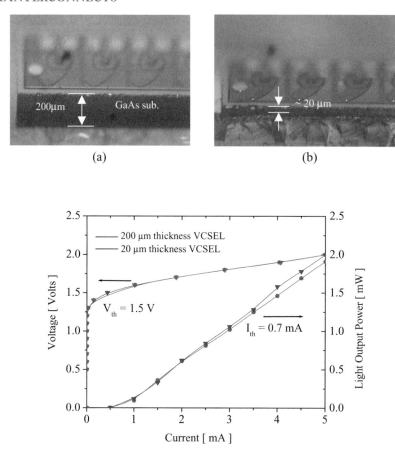

FIGURE 8.2: (a) 200 μm thick VCSEL array (before substrate thinning), (b) 20 μm thick VCSEL array (after substrate thinning), (c) electrical and optical properties of the VCSEL (eye-diagram measured at 5 mA/2 V bias condition)

pattern generator provides the modulation by generating a $2^{23} - 1$ pseudorandom bit sequence (PRBS) nonreturn-to-zero (NRZ) pattern at 10 Gbps. The emitted light is collected by a multimode fiber (MMF, $\varphi = 62.5\mu$m) and detected by a 21 GHz bandwidth photo-receiver (Newport, D-15). The measured 10 Gbps eye-diagram shows reasonable opening.

Due to the self-heating effect of the VCSEL, the temperature of the active region rises relative to the heat sink. To achieve the maximum VCSEL reliability in the fully embedded structure, it is imperative to control the VCSEL temperature during operation. The ratio of the temperature rise (ΔT) to the net dissipation power (ΔP_{diss}) is defined as the thermal resistance (R_{th}). The thermal resistance (R_{th}) evaluation of the VCSEL as a function of the substrate

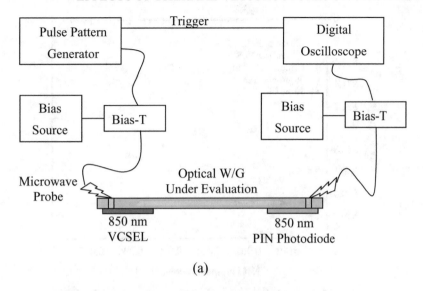

(a)

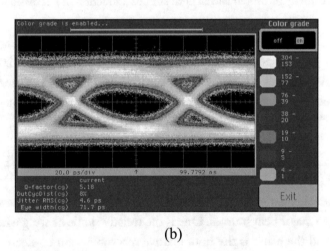

(b)

FIGURE 8.3: (a) Schematic diagram of the speed measurement setup, (b) 10 Gbps eye-diagram measured at 5 mA/2 V bias condition

thickness is performed by measuring the wavelength shift with both the temperature $(\Delta\lambda/\Delta T)$ and the net dissipated power $(\Delta\lambda/\Delta P_{\text{diss}})$. The thermal resistance is given by

$$R_{\text{th}} = \Delta T/\Delta P_{\text{diss}} = (\Delta\lambda/\Delta P_{\text{diss}})/(\Delta T/\Delta\lambda) \qquad (8.1)$$

where ΔT is the temperature rise in the active region, ΔP_{diss} is the change of electrical power dissipated in the VCSEL, and $\Delta\lambda$ is the wavelength shift. A VCSEL array is mounted on a thermoelectric cooler (TEC) to precisely control the substrate temperature at $25°C \pm 1$. The

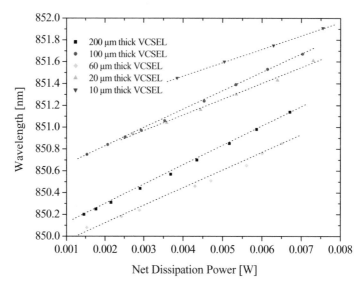

FIGURE 8.4: Experimentally measured wavelength shift as a function of net dissipated power

measured wavelength temperature variation $(\Delta\lambda/\Delta T)$ is 0.066 nm/°C. In Fig. 8.4, the slopes of the wavelength shift as a function of net dissipated power $(\Delta\lambda/\Delta P_{\mathrm{diss}})$ for 200 μm, 100 μm, 60 μm, 20 μm, and 10 μm thick VCSELs are 188 nm/W, 169 nm/W, 153 nm/W, 144 nm/W, and 120 nm/W, respectively. Experimentally measured thermal resistance values for each device are 2848°C/W, 2560°C/W, 2326°C/W, 2181°C/W, and 1830°C/W.

8.3   NUMERICAL MODELING OF SELF-HEATING EFFECT FOR A THIN FILM VCSEL

Several studies on the self-heat generation mechanisms inside the VCSEL have been reported [68, 69]. There are two major heat sources. One is the resistive joule heating when current flows through the DBRs, and the other is the nonradiative recombination of electrons and holes in the active region. To simulate inside the VCSEL temperature rise, the heat source distributions inside the VCSEL have to be determined. In this chapter, a 2D thermal-electric direct coupled-field analysis module of ANSYS software is used to calculate the temperature distribution due to the joule heating in the DBRs and the nonradiative recombination in the active region. In this simulation, we assumed that the VCSEL is azimuthal symmetry. To consider interface and boundary scattering effects of phonons and electrons, anisotropic material properties are used in the DBR region. Table 8.1 lists the electrical and thermal properties used in this simulation.

The thermal-electric simulation is limited to the steady-state analysis. First, the electric field analysis is performed to calculate the current density distributions inside the VCSEL near the active region. Then, the thermal analysis is conducted to calculate the joule heating and the

TABLE 8.1: Materials Properties and Physical Dimensions Used in Simulation.

		THERMAL CONDUCTIVITY (W/μm K)		ELECTRICAL CONDUCTIVITY ($\Omega^{-1}\mu m^{-1}$)	THICKNESS (μm)
p-DBR	k_r/k_z	$1.2 \times 10^{-5}/ 1.0 \times 10^{-5}$	σ_r/σ_z	$2.016 \times 10^{-3}/ 1.5 \times 10^{-5}$	3.30
n-DBR	k_r/k_z	$1.2 \times 10^{-5}/ 1.0 \times 10^{-5}$	σ_r/σ_z	$4.03 \times 10^{-2} / 2.85 \times 10^{-4}$	3.497
Substrate	$k_r = k_z$	4.5×10^{-5}	$\sigma_r = \sigma_z$	3.3×10^{-2}	10–200
Au	$k_r = k_z$	3.1×10^{-4}	$\sigma_r = \sigma_z$	45.4	0.2
Copper	$k_r = k_z$	3.86×10^{-4}	$\sigma_r = \sigma_z$	58.8	0–200
Polymer	$k_r = k_z$	2.0×10^{-7}	$\sigma_r = \sigma_z$	10^{-18}	30–100

temperature distributions inside the VCSEL. The local heat generation rate due to the joule heating is calculated by

$$q = \sigma_z \left(\frac{\partial V}{\partial z} \right)^2 + \sigma_r \left(\frac{\partial V}{\partial r} \right)^2 \qquad (8.2)$$

where σ_z and σ_r are the electrical conductivity in the z-axial and the radial directions, respectively [68]. A 2-V potential drop boundary condition, equivalent to our experimental work, is used for the electric field analysis. The steady-state heat conduction equation for the axial symmetric structure is governed by

$$k_z \frac{\partial^2 T}{\partial z^2} + k_r \frac{\partial^2 T}{\partial r^2} + q(z, r) = 0 \qquad (8.3)$$

where q is the local heat generation rate. k_z and k_r are the thermal conductivity along the z-axial and the radial directions, respectively [72].

Figure 8.5(a) shows a mesh generated 2D modeling structure of the VCSEL near the active region. Both the Si-doped n-type DBR and the C-doped p-type DBR consist of stacks of quarter-wavelength layers of GaAs and $Al_{0.9}Ga_{0.1}As$. The current aperture and the active region are located between the two reflectors. Symmetric and adiabatic boundary conditions are applied on the sidewall and the top surface of the modeling structure. The boundary condition for the bottom surface of the Cu plate is 25°C. The radial distributions of temperature rise and heat generation density near the active region are shown in Fig. 8.5(c). According to the electrical potential distribution inside the VCSEL, most of voltage drop occurs across the p-type DBR. As shown in Fig. 8.5(b), an abrupt voltage drop near the edge of the active region indicates a strong current concentration when a great portion of current converges into the

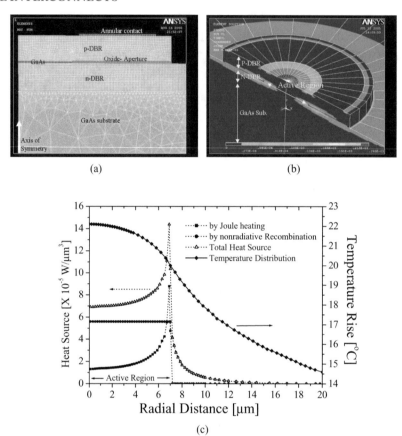

FIGURE 8.5: (a) 2D symmetric modeling of the VCSEL and mesh structure near the active region, (b) current density distribution inside the VCSEL near the active region, (c) heat source and temperature rise distributions of the 200 μm thick VCSEL near the active region (5 mA/2 V bias condition)

active region. This current concentration gives rise to a heat source spike at the edge of the active region as shown in Fig. 8.5(c). However, this spike shaped heat source distribution does not cause a local temperature peak distribution. The temperature distribution peaks along the optical axis. These observations are consistent with the results of Chen [68].

The calculated and experimentally measured thermal resistances are shown in Fig. 8.6 with ambient temperature set at 25°C. The calculated thermal resistances as a function of the substrate thickness of the VCSEL are matched well with the experimentally measured results. This result shows that the simulation models for this study are properly carried out and thinned VCSEL has an exclusive advantage of heat management due to the reduction of the thermal resistance. The thermal resistance of a 10 μm thick VCSEL is 40% lower than that of a 200 μm thick VCSEL. The calculated active region temperatures for the 200 μm and 10 μm thick VCSELs are 47.16°C and 38.9°C, respectively.

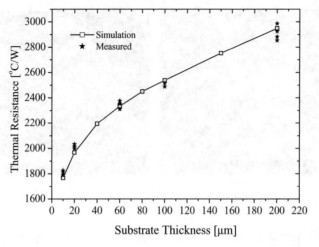

FIGURE 8.6: Experimentally measured and simulation results of the thermal resistances as a function of the substrate thickness of the thin film VCSEL

8.4   THERMAL-VIA STRUCTURES FOR THE FULLY EMBEDDED THIN FILM VCSEL

Thermally, via plays a significant roll in locally enhancing the heat conduction through the board because the thermal conductivity of copper is 1200 times that of common dielectric materials [73]. On the backside of the fully embedded thin film VCSEL, a thermal blind-via, where it enters one side and stop at an internal layer, will be applied as a heat sink structure. Figures 8.7(a) and (b) show simulation results of temperature distribution inside the VCSEL with two commonly employed thermal-via structures, the closed blind-via (copper filled inside the via hole (a)) and the open blind-via (30 μm thick copper electroplated inside the via hole (b)), respectively. In this simulation model, we assumed that the top of the VCSEL is covered by a polymer layer, which is the waveguiding layer as shown in [65, 74].

Figure 8.7(c) shows theoretically determined thermal resistances of the fully embedded thin film VCSEL with different thermal-via structures. Following the previous results (Fig. 8.6), the compatible thermal resistance for the fully embedded VCSEL structure should be in the same range of the 200 μm thick VCSEL with 25°C substrate cooling condition. Therefore, in this study, the target thermal resistance is determined to be in the range of 2800–3000°C/W. In the case of the open blind-via structures (straight lines in Fig. 8.7(c)) with an aspect ratio of 1 and 0.5, the effective substrate thickness of the thin film VCSEL is estimated in the range of 12–22 μm and 29–46 μm, respectively. Also, in the case of the closed blind-via structures (dot lines in Fig. 8.7(c)) with an aspect ratio of 1 and 0.5, the fully embedded structure compatible substrate thickness of the VCSEL is in the range of 44–60 μm and 53–72 μm, respectively. Figure 8.8 shows a via hole structure on the optical film fabricated with a diameter of 200 μm and an aspect ratio of 0.5. The copper pillar inside via hole is clearly shown. Such a thermal via

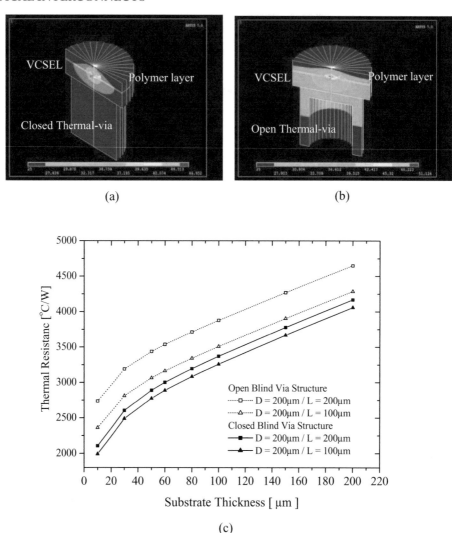

(a)   (b)

(c)

FIGURE 8.7: (a) 50 μm thick VCSEL with a backside closed blind-via structure, (b) 50 μm thick VCSEL with a backside open blind-via structure, (c) simulation results of thermal resistance as a function of the substrate thickness of the thin film VCSEL with different thermal-via structures

will be integrated according to the scheme shown in Fig. 8.1. Further results will be presented in the future.

8.5   CONCLUSION

The thermal resistances of the substrate thinned VCSEL are experimentally measured and calculated using the 2D thermal-electric coupled field analysis method. Both the joule heating and the nonradiative recombination effects are considered to calculate the heat source and the

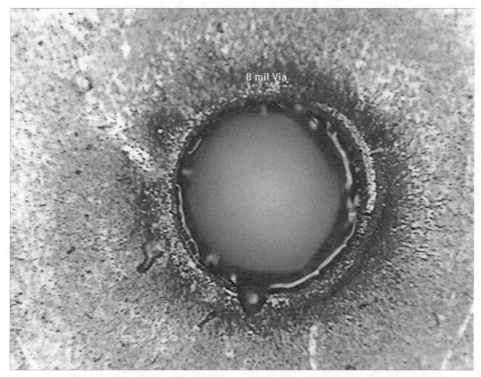

FIGURE 8.8: Fabricated via hole on the optical film layer ($D = 200$ μm, aspect ratio $= 0.5$)

temperature rise distributions inside the VCSEL near the active region. The thermal-electric analysis results matched well with experimental data. Simulation models for this study are properly carried out, and a thinned VCSEL had an exclusive advantage of the heat management. The thermal resistance of a 10 μm thick VCSEL is 40% lower than that of a 200 μm thick VCSEL. The calculated active region temperatures for 200 μm and 10 μm thick VCSELs are 47.16°C and 38.9°C, respectively. According to the theoretical analysis of the thermal-via structures for the fully embedded thin film VCSEL, the calculated thin film VCSEL thickness compatible with the fully embedded board-level optical interconnection system depends on the thermal-via structures. In the case of the closed thermal-via structures, the substrate thickness of the VCSEL in the range of 44–72 μm is conformable in the fully embedded board-level optical interconnection system.

8.6   ACKNOWLEDGMENTS
This research is supported by Darpa, ONR and Texas ATP Program.

Bibliography

[1] D. P. Seraphim and D. E. Barr, "Interconnect and packaging technology in the 90s," *Proc. SPIE.*, vol. 1390, p. 39, 1990.

[2] R. O. C. Neugerbauer, R. A. Fillion, and T. R. Haller, "Multichip module designs for high performance applications," *Multichip Modules, Compendium of 1989 Papers*, International Electronic Packaging Society, p. 149.

[3] SIA, *The National Technology Roadmap for Semiconductors—Technology Needs*. Semiconductor Industry Association, 1999–2000.

[4] N. Cravotta, "Wrestlemania: keeping high-speed-backplane design under control," *EDN*, pp. 36–46, Aug. 2002.

[5] L. W. Schaper, "Flex and the interconnected mesh power system," Chapter 12, in *Foldable Flex and Thinned Silicon Multichip Packaging Technology*, Boston: Kluwer Academic, 2003.

[6] R. C. Walker, K. C. Hsieh, T. A. Knotts, and C. S. Yen, "A 10 Gb/s Si-bipolar TX/RX chipset for computer data transmission," in *Proc. IEEE Int. Solid-State Circuits Conf.*, Feb. 1998, p. 302.

[7] A. F. Levi, "Optical interconnects in system," *Proc. IEEE*, vol. 88, p. 750, 2000. doi:10.1109/5.867688

[8] M. Gruber, S. Sinzinger, and J. Jahns, "Optoelectronic multichip module based on planar-integrated free-space optics," in *Proc. SPIE, Optics Comput. 2000*, vol. 4089.

[9] G. Kim and R. T. Chen, "Three-dimensionally interconnected multi-bus-line bidirectional optical backplane," *Opt. Eng.*, vol. 38, p. 9, 1999.doi:10.1117/1.602074

[10] E. Griese, "Parallel optical interconnects for high performance printed circuit board," in *Proc. 6th Int. Conf.*, 1999, p. 173.

[11] Y. Ishii, S. Koike, Y. Aria, and Y. Ando, "SMT-compatible optical I/O chip packaging for chip-level optical interconnects," *Electron. Components Technol. Conf.*, 2001.

[12] D. Krabe, F. Ebling, N. Arndt-Staufenbiel, G. Lang, and W. Scheel, "New technology for electrical/optical systems on module and board level: The EOCB approach," in *Proc. 50th Electron. Components Technol. Conf.*, Las Vegas, NV, May 2000, p. 970.

[13] K. Iga, F. Koyama, and S. Kinoshita, "Surface emitting semiconductor lasers," *IEEE J. Quant. Elect.*, vol. 24, 1988.

[14] H. A. Macleod, "*Thin-Film Optical Filters*," 2nd ed. Bristol: Adam Hilger, 1986.

[15] D. E. Aspens, S. M. Kelso, R. A. Logan, and R. Bhat, "Optical properties of Al$_x$Ga$_{1-x}$As," *J. Appl. Phys.*, vol. 60(2), p. 754, 1986.doi:10.1063/1.337426

[16] W. T. Tsang, "Self-terminating thermal oxidation of AlAs epilayers grown on GaAs by molecular beam epitaxy," *Appl. Phys. Lett.*, vol. 33, p. 426, 1978.

[17] J. M. Dallesasse, N. Holonyak, Jr., A. R. Sugg, T. A. Richard, and N. El-Zein, "Hydrolyzation oxidation of Al$_x$Ga$_{1-x}$As-AlAs-GaAs quantum well heterostructures and superlattices," *Appl. Phys. Lett.*, vol. 57, p. 2884, 1990.

[18] G. Stareev, "Formation of extremely low resistance Ti/Pt/Au ohmic contacts to p-GaAs," *App. Phys. Lett.*, vol. 62(22), p. 2801, 1993.doi:10.1063/1.109214

[19] J. W. Balde, Ed., Chapter 5 in *Foldable Flex and Thinned Silicon Multichip Packaging Technology*, Boston: Kluwer Academic, 2003.

[20] M. Feil, C. Landesberger, A. Klumpp, and E. Hacker, "Method of subdividing a wafer," Patent Application WO 01/03180 A1.

[21] E. Yablovitch, T. J. Gmitter, J. P. Harbison, and R. J. Bhat, "Extreme selectivity in the lift-off of epitaxial GaAs films," *Appl. Phys. Lett.*, vol. 51, p. 2222, 1987.doi:10.1063/1.98946

[22] Y. Sasaki, T. Katayama, T. Koishi, K. Shibahara, S. Yokohama, S. Miyazaki, and M. Hirose, "High-speed GaAs epitaxial lift-off and bonding with high alignment accuracy using a sapphire plate," *J. Electrochem. Soc.*, vol. 146(2), p. 710, 1999. doi:10.1149/1.1391668

[23] O. Vedier, N. Jokerst, and R. P. Leavitt, "Thin-film inverted MSM photodetectors," *IEEE Photon. Tech. Lett.*, vol. 8(2), p. 266, 1996.doi:10.1109/68.484262

[24] B. D. Dingle, M. B. Spitzer, R. W. McClelland, J. C. C. Fan, and P. M. Zavracky, "Monolithic integration of a light emitting diode array and a silicon circuit using transfer process," *Appl. Phys. Lett.*, vol. 62(22), p. 2760, 1993.doi:10.1063/1.109252

[25] A. J. Tsao, Dissertation, University of Texas at Austin, 1993.

[26] C. Juang, K. J. Kuhn, and R. B. Darling, "Selective etching of GaAs and Al0.3Ga0.7As with citric acid/hydrogen peroxide solutions," *J. Vac. Sci. Technol.*, vol. B8, 1990.

[27] J.-L. Lee, F.-A. Moon, J.-W. Oh, S. W. Ryu, and H. M. Yoo, "Selective wet etching of GaAs on Al0.24Ga0.76As for GaAs/Al0.24Ga0.76As/In0.22Ga0.78As PHEMT," *Elecron. Lett.*, vol. 36, 2000.

[28] H. Fathollahnejad, D. L. Mathine, R. Droopad, G. N. Maracas, and S. Daryanani, "Vertical-cavity surface emitting lasers integrated onto silicon substrates by PdGe contacts", *Electron. Lett.*, vol. 30(15), 1994.

[29] A. Ghiti, M. Silver, and E. P. O'Reilly, "Low threshold current and high differential gain in ideal tensile and compressive strained quantum well lasers", *J. Appl. Phys.*, vol. 35, p. 4626, 1992.doi:10.1063/1.350766

[30] L. A. Coldren and S. W. Corzin, Chapter 5 in *Diode Lasers and Photonic Integrated Circuits*, New York: Wiley, 1995.

[31] Y. C. Lee, S. E. Swirhun, W. S. Fu, T. A. Keyser, J. L. Jewell, and W. E. Quinn, "Thermal management of VCSEL-based optoelectronic modules," *IEEE Trans. Compon. Packag. Manuf. Technol.*, vol. 19(3), pp. 540–547, 1996.doi:10.1109/96.533893

[32] R. Pu, C. W. Wilmsen, K. M. Geib, and K. D. Choquette, "Thermal of VCSEL's bonded to integrated circuits," *IEEE Photon. Technol. Lett.*, vol. 11(12), pp. 1554–1556, 1999. doi:10.1109/68.806844

[33] T. Yao, "Thermal properties of AlAs/GaAs superlattices," *Appl. Phys. Lett.*, vol. 51, p. 1798, 1987.doi:10.1063/1.98526

[34] J. Piprek, T. Tröger, B. Schröter, J. Kolodzey, and C. S. Ih, "Thermal conductivity reduction in GaAs–AlAs distributed bragg reflectors," *IEEE Photon. Technol. Lett.*, vol. 10, 1998.

[35] T. Suhara and H. Nishihara, "Integrated optics components and devices using periodic structures," *IEEE J. Quantum Electron.*, vol. QE-22, p. 845, 1996.

[36] D. Brundrett, E. Glystis, and T. Gaylord, "Homogeneous layer models for high-spatial-frequency dielectric surface-relief gratings," *Appl. Opt.*, vol. 33, p. 2695, 1994.

[37] D. Y. Kim, S. K. Tripathy, L. Li, and J. Kumar, "Laser-induced holographic surface relief gratings on nonlinear optical polymer films," *Appl. Phys. Lett.*, vol. 66, p. 1166, 1995. doi:10.1063/1.113845

[38] R. T. Chen, F. Li, M. Dubinovsky, and O. Ershov, "Si-based surface-relief polygonal gratings for 1-to-many wafer-scale optical clock signal distribution," *IEEE Photon. Technol. Lett.*, vol. 8, 1996.

[39] R. K. Kostuk, J. W. Goodman, and L. Hesselink, "Design considerations for holographic optical interconnects," *Appl. Opt.*, vol. 26, p. 3947, 1987.

[40] L. Eldada and J. T. Yardly, "Integration of polymeric micro-optical elements with planar waveguiding circuits," *Proc. SPIE*, vol. 3289, p. 122, 1998.doi:full_text

[41] R. T. Chen, L. Lin, C. C. Choi, Y. Liu, B. Bihari, L. Wu, R. Wickman, B. Picor, M. K. Hibbs-Brenner, J. Bristow, and Y. S. Liu, "Fully embedded board-level guided-wave optoelectronic interconnects," *Proc. IEEE*, vol. 88(6), pp. 780–793, 2000. doi:10.1109/5.867692

[42] M. Kagami, A. Kawasaki, and H. Ito, "A polymer optical waveguide with out of plane branching mirrors for surface-normal optical interconnections," *J. Lightwave Technol.*, vol. 19(12), 2001.

[43] H. Terui, M. Shimokozono, M. Yanagisawa, T. Hashimoto, Y. Yamashida, and M. Horiguchi, "Hybrid integration of eight channel PD-array on silica based PLC using micro-mirror fabrication technique," *Elect. Lett.*, vol. 32(18), 1996.

[44] Hikita, R. Yoshimura, M. Usui, S. Tomaru, and S. Imamura, "Polymeric optical waveguides for optical interconnections," *Thin Solid Films*, vol. 331, p. 303, 1998. doi:10.1016/S0040-6090(98)00935-3

[45] K. D. Choquette and K. M. Geib, "Fabrication and performance of vertical cavity surface emitting lasers" Chapter 5 in *Vertical Cavity Surface Emitting Lasers*, C. Wilmsen, H. Temkin, and L. A. Coldren, Eds., Cambridge, UK: Cambridge University Press, 1999.

[46] H. J. Unold, S. W. Z. Mahmoud, R. Jager, M. Kicherer, M. C. Riedl, and K. J. Ebeling, "Improving single-mode VCSEL performance by introducing a long monolithic cavity," *IEEE Photon. Technol. Lett.*, vol. 12(8), p. 939, 2000.doi:10.1109/68.867967

[47] N. Nishiyama, M. Arai, S. Shinada, K. Suzuki, F. Koyama, and K. Iga, "Multi-oxide layer structure for single-mode operation in vertical-cavity surface-emitting lasers," *IEEE Photon. Technol. Lett.*, vol. 12(6), p. 606, 2000.doi:10.1109/68.849058

[48] D.-S. Song, S.-H. Kim, H.-G. Park, C.-K. Kim, and Y.g H. Lee, "Single-mode photonic-crystal vertical cavity surface emitting laser," *CLEO '02. Tech. Digest*, vol. 1, p. 293, 2002.

[49] B. E. A. Saleh and M. C. Teich, Chapter 3 in *Fundamentals of Photonics*, New York: Wiley, 1991.

[50] L. Ristic, "Sensor technology and devices", Chapter 3 in *Bulk Micro Machining Technology*, L. Ristic, H. Hughes, and F. Shemansky, Eds., Boston, London: Artech House, 1994.

[51] H. Seidel, "The mechanism of anisotropic, electrochemical silicon etching in alkaline solutions," *IEEE Solid-State Sensor and Actuator Workshop*, Hilton Head Island, SC, 1990, p. 86.

[52] D. Corning, "Product information sheet," Sylgard 184.

[53] Y. Xia and G. M. Whitesides, "Soft lithography," *Annu. Rev. Mater. Sci.*, 1998.

[54] Microchem, Micorchem product references, http://www.microchem.com/products/pdf/SU8_2035-2000.pdf

[55] Y. J. Chuang, F. G. Tseng, and W. K. Lin, "Reduction of diffraction effect of UV exposure on SU-8 negative hick photoresist by air gap elimination," *Microsyst. Technol.*, vol. 8, p. 308, 2002.doi:10.1007/s00542-002-0176-8

[56] G. T. A. Kovacs, Chapter 3 in *Micromachined Tranducers*, Boston: McGraw-Hill, 1998.

[57] P. D. Curtis, S. Iezeliel, R. E. Miles, and C. R. Pescod, "Preliminary investigations into SU-8 as a material for integrated all-optical microwave filters," *High Frequency Postgraduate Student Colloquium, 2000*, 7–8 Sept. 2000, p. 116.

[58] W. H. Wong, J. Zhou, and E. Y. B. Pun, "Low-loss polymeric optical waveguides using electron-beam direct writing," *Appl. Phys. Lett.*, vol. 78, p. 2110, 2001. doi:10.1063/1.1361287

[59] E. Yablovitch, T. Sands, D. M. Hwang, I. Schnitzer, and T. J. Gmitter, "Van der Waals bonding of GaAs on Pd leads to a permanent, solid-phase-topotaxial, metallurgical bond," *Appl. Phy. Lett.*, vol. 59(24), p. 3159, 1991.doi:10.1063/1.105771

[60] International SEMATECH, *The National Technology Roadmap for Semiconductor (ITRS)-Technology Needs*, Semiconductor Industry Association, 2004.

[61] M. Forbes, J. Gourlay, and M. Desmulliez, "Optically interconnected electronic chips: a tutorial & review of the technology," *Electron. Commun. Eng. J.*, pp. 221–232, 2001. doi:10.1049/ecej:20010506

[62] Y. Ishii, N. Tanaka, T. Sakamoto, and H. Takahara, "Fully SMT-compatible optical I/O package with microlens array interface," *J. Lightwave Technol.*, vol. 21(1), pp. 275–280, 2003. doi:10.1109/JLT.2003.808611

[63] P. V. Daele, P. Geerinck, G. V. Steenberge, and S. V. Put, "Optical interconnections on PCB's: a killer application for VCSEL's," *Proc. SPIE*, vol. 4942, pp. 269–281, 2003. doi:full_text

[64] C. C. Choi, L. Lin, Y. Liu, J. H. Choi, W. Li, D. Haas, J. Magera, and R. T. Chen, "Flexible optical waveguide film fabrications and optoelectronic devices integration for fully embedded board level optical interconnects," *J. Lightwave Technol.*, vol. 22(6), pp. 2168–2176, 2004. doi:10.1109/JLT.2004.833815

[65] J. H. Choi, W. Li, X. Wang, D. Haas, J. Magera, and R. T. Chen, "Performance evaluation of fully embedded board level optical interconnections," *IEEE-LEOS Summer Topicals Meeting Series-Optical Interconnects & VLSI Photonics*, pp. 9–10, 2004.

[66] Ray T. Chen, "Packaging enhanced board level opto-electronic interconnects," US Patent No. 6243509.

[67] M. Fukuda, *Reliability and Degradation of Semiconductor Lasers and LEDs*, Boston: Artech House, 1991.

[68] G. Chen, "A comprehensive study on the thermal characteristics of vertical-cavity surface-emitting lasers," *J. Appl. Phys.*, vol. 77(9), pp. 4251–4258, 1995. doi:10.1063/1.359481

[69] Y. Liu, W. C. Ng, K. D. Choquette, and K. Hess, "Numerical investigation of self-heating effects of oxide-confined vertical-cavity surface-emitting lasers," *IEEE J. Quantum Electron.*, vol. 41(1), pp. 15–25, 2005. doi:10.1109/JQE.2004.839239

[70] A. V. Krishnamoorthy, *et al.*, "16X16 VCSEL array flip-chip bonded to CMOS VLSI circuit," *IEEE Photon. Technol. Lett.*, vol. 12(8), pp. 1073–1075, 2000. doi:10.1109/68.868012

[71] C. K. Lin, S. W. Ryu, and P. D. Dapkus, "High-performance wafer-bonded bottom-emitting 850 nm VCSEL's on undoped GaP and saphire substrates," *IEEE Photon. Technol. Lett.*, vol. 11(12), pp. 1542–1544, 1999. doi:10.1109/68.806840

[72] F. P. Incropera and D. P. Dewitt, *Introduction to Heat Transfer*, New York: Wiley.

[73] A. M. Lush, "Modelling heat conduction in printed circuit boards using finite element analysis," *Electron. Cooling*, vol. 10(2), 2004.

[74] L. Wang, W. Jiang, X. Wang, J. H. Choi, H. Bi, and R. T. Chen, "45° polymer-based total internal reflection coupling mirrors for fully embedded intraboard guided wave optical interconnects," *J. Appl. Phys. Lett.*, 2005.

Printed in the United States
by Baker & Taylor Publisher Services